Skills for a Scientific Life

Skills for a Scientific Life

JOHN R. HELLIWELL

CRC Press
Taylor & Francis Group
Boca Raton London New York

CRC Press is an imprint of the
Taylor & Francis Group, an **informa** business

CRC Press
Taylor & Francis Group
6000 Broken Sound Parkway NW, Suite 300
Boca Raton, FL 33487-2742

Library of Congress Cataloging-in-Publication Data

Names: Helliwell, John R.
Title: Skills for a scientific life / John R. Helliwell.
Description: New York : CRC Press, [2017] | Includes bibliographical references and index.
Identifiers: LCCN 2016028050 | ISBN 9781498768757 (hardback)
Subjects: LCSH: Science--Vocational guidance. | Science--Social aspects. | Scientists. | Teams in the workplace.
Classification: LCC Q180.55.V6 H45 2017 | DDC 502.3--dc23
LC record available at https://lccn.loc.gov/2016028050

Visit the Taylor & Francis Web site at
http://www.taylorandfrancis.com

and the CRC Press Web site at
http://www.crcpress.com

To my mother and father; to my wife Madeleine and our son James and daughter Katherine.

In loving memory of Nick Helliwell (1983–2011).

Contents

SECTION I   *Introduction: How Do You Know You Are Suited to Be a Scientist*

SECTION II   *Skills for a Better Researcher*

SECTION III   Being a Good Science Research Citizen

SECTION IV   Skills for Being an Educator

SECTION VIII   Appendices

Preface

Being, or wanting to become, a scientist obviously requires academic training in the science subjects. To succeed as a research scientist and/or an educator requires specific skills of research article writing, grant writing, refereeing and so on, as well as teaching undergraduates and supervising postgraduates if you are an academic. Less immediately obvious are that general skills are also necessary to be an effective scientist. In addition, how one approaches running one's own research laboratory, or how one conducts oneself as a member of a research laboratory led by someone else, can have profound consequences on issues such as gaining (or losing) one's own research skills, societal impacts (such as via press releases) and gender equality. Then there is the possible lure of politics, but how to avoid such is a skill in itself. Another skill is how one judges others, whether refereeing their grant proposals or submitted articles; those roles are usually under the cloak of anonymity. A very public task is to review books; the book review presents much that is of scientific interest, and one has to be an expert to review an academic book. But being fair to the book author and the potential buyer of a book in a book review is an important and yet delicate line to tread. Also, you will meet many scientists throughout your career, some not always with a rigorous view of ethics, so be rightly wary before proposing to a possible collaborator!

I have led a successful research career and trained successful students. I have worked as an academic in physics and chemistry departments and taught biochemists and molecular biologists as a guest lecturer. I have degrees in physics and molecular biophysics and been a professor of chemistry. Most unusually, I have worked as well in the scientific civil service up to department head level (staff composed of 240 people). This latter presented me with a wide range of general management training that complemented my broad-ranging academic training. At the University of Manchester, I have served as gender equality champion, leading the team in transforming my academic department to a Bronze Athena Science Women's Academic Network [SWAN] awardee and then making the successful proposal to gain a Silver Athena SWAN Award. Athena SWAN is the initiative of the UK government (http://www.ecu.ac.uk/equality-charters/athena-swan/) established in 2005 to 'encourage and recognise commitment to advancing the careers of women in science, technology, engineering, maths and medicine employment in higher education and research'. I also served as a senior mentor for new academics in my department. Finally, I have served the University of Manchester as a mentor in its Manchester Gold scheme mentoring academic staff in the Faculties of Science and Engineering and in Clinical Sciences and Medicine. I have served as a science research editor up to editor-in-chief level; this is where one applies one's science research experience and knowledge, working with the administration office managing editor. I have judged the acceptability of around 1000 submitted research articles and around 100 research reviews and book reviews. The topics of this book are underpinned by all this detailed and wide-ranging experience, including my own case studies throughout my career.

This book then provides my insights into the skills needed by a scientist as a researcher and an educator. This book will be valuable to scientists at all stages of their careers. It will also be valuable to children choosing between science or arts and humanities subjects, and likewise their advisors and their parents; this important decision is taken by schoolchildren in England at the age of 15 years.

I have received prestigious prizes for my research in the United States and in Europe. I have been privileged to spend more than 25 years at the University of Manchester, one of the most successful postcodes 'M13 9PL' for Nobel Prizes in chemistry and physics! I hope my book will assist your own science career aspirations, and through its case studies, and the quotations I have chosen for you, may it not only serve as a guide but also give you some entertainment. As a guide, via a quote, I have selected the one from Jesse Owens, with him being widely known as a four-time Olympic gold medallist at the Berlin Olympics in 1936, including three world records. But not only that, his respect for others in that quotation also shines as a beacon, which surely kept his peace of mind. As you strive for excellence via your own skills, and success, never lose your conscience or peace of mind; that is, overall, the most important aspect of your career in science.

Choosing the chapter and section titles was one of the most important steps in the conception of this book; I have enthusiastically thanked the referees of my book proposal in their help on this. In my chapter headings, I have opted for the rhetorical how-to approach; *rhetorical* means I have not included a question mark. These are explained, for example, in a nice piece in Wikipedia: 'The effectiveness of rhetorical questions in argument comes from their dramatic quality. They suggest dialogue, especially when the speaker both asks and answers them himself, as if he were playing two parts on the stage'. In my writing style, I have chosen a conversational one, addressing you wherever possible. As a long-standing mentor, I feel that this writing style is as if I am the mentor with you as the mentee. It also has the advantage of avoiding that problem of using *his* and the more modern switch to *his/her* employed by some authors in their coaching writings. You can obviously read the book from start to finish, but in your work and career urgent situations may arise where for you to dip into the appropriate chapter would surely be advantageous, and possible, yes absolutely.

John R. Helliwell
Emeritus Professor of Chemistry, University of Manchester;
DSc Physics, University of York

Acknowledgements

I thank the Universities of York, Oxford, Keele and Manchester (since 1989); the Synchrotron Radiation Source at Daresbury Laboratory (1976–2008); the Diamond Light Source at the Rutherford Appleton Laboratory; the European Synchrotron Radiation Facility and the Institut Laue–Langevin neutron source in Grenoble, for each providing me with splendid environments in which to study and undertake my scientific research and development activities. I thank all my students and research colleagues for their collaborations. I thank all the research funding agencies that supported my research, namely the UK's: Science and Engineering Research Council, then the Biotechnology and Biological Sciences Research Council, the Engineering and Physical Sciences Research Council and the Medical Research Council; the Wellcome Trust; the Leverhulme Trust; the Nuffield Foundation; the Royal Society; and the British Council. I thank the international funding agencies: the Swedish Research Council, the Swedish Hasselblad Foundation, the European Space Agency, the North Atlantic Treaty Organisation, and finally, on numerous occasions, the European Community Framework Programmes. I thank the International Union of Crystallography for being such an effective global organisation supporting my science. I also thank all those conference organisers who provided a framework for us as scientists to meet and confer and the scientific organisations that supported those conferences and furthermore enrich our professional lives. Without all this making a fantastic environment, and my wonderful colleagues and students, from the UK and around the world, this book would never have happened. A hearty thank you to all!

I thank Hilary LaFoe at CRC Press for her wise advice as my publisher. I thank Professor Moreton Moore as a nonanonymous referee for his insightful comments on my book proposal and likewise the anonymous referee. These three people encouraged me in the writing of a more general book on science and from which my book title emerged and with its expanded contents list.

The books in the Bibliography gave me many wider insights. I record here my hearty acknowledgement to these book authors (full details are in the Bibliography): Peter Medawar (1915–1987), Edward Wilson, Finlay MacRitchie, Joëlle Fanghanel, Gerald Kaufmann and Erich Fromm (1900–1980).

I especially thank my wife, also a scientist, Dr Madeleine Helliwell, for her love and companionship as well as research collaborations these last 40 years.

I would like to quote one of the referees of my book proposal, and whose wise advice I have kept in mind during my writing of this book:

> This book proposal's headings focus on the formal side of training in various disciplines. The primary goal should be to make the readers and future scientists enthusiastic and be inspired by some of the giants on whose shoulders we stand and let them guide us. I cannot think the formal aspects would ever make a student an excellent scientist. But once you get excited you should consider the formal side. To become a good scientist you need to identify the ingredients that form a strong intellectual infrastructure. This is a prime point to become a good scientist. The UK's Medical Research Council Laboratory of Molecular Biology in Cambridge is perhaps the most

outstanding example worldwide; the harvest of Nobel Prizes to full and temporary employees of this laboratory is unmatched. The intellectual infrastructure should be highlighted. When you ask people from this lab why they have such a success the answer is unanimously: THE CANTEEN! The daily discussions of scientists from a wide range of fields cannot fail to discuss unsolved problems that are worthwhile and mature for an attempt. Too many labs are essentially lacking an intellectual infrastructure. Being, or wanting to become, a scientist you need to be curious. Furthermore, as Francis Crick put it: 'If you ask big questions, you get big answers'. Then you obviously require academic training in the science subjects. Secondary to asking big questions to succeed as a research scientist and educator requires specific skills of research article writing, grant writing, refereeing and so on as well as teaching undergraduates and supervising postgraduates.

I thank the following people and or companies or organisations for allowing me to reproduce text extracts or figures: Professor Jennifer Doudna, Dr Caterina Biscari, Professor Nancy Rothwell, Taylor and Francis, L'Oreal Foundation, UNESCO and the National Physical Laboratory and the International Union of Crystallography.

Finally, a number of colleagues kindly gave their time and thoughts on reading the early drafts of the chapters. I heartily thank the following people: Dr Colin Bulpitt, Dr Caterina Biscari, Peter Strickland, Professor Alessia Bacchi and Professor Leann Tilley. The following colleagues kindly accepted my request to comment on the whole draft: Emeritus Professor Moreton Moore, Dr Titus Boggon, Dr Michele Cianci, Dr Charlie Bond, Dr Andy Thompson, Dr Eddie Snell, Katrine Bazeley, Dr Petra Bombicz, Dr Jill Barber, Dr Briony Yorke and Dr Matthew Blakeley.

I chose these people because I knew them and that they would be helpful, but I also chose them to be representative in their own way of a wide range of countries (United Kingdom, United States, Australia, Germany, France, Spain, Italy and Hungary).

Any mistakes or oddities of perception of the scientific life that might remain are of course my own.

Author

John R. Helliwell is an emeritus professor of structural chemistry, University of Manchester. He has a DPhil degree in molecular biophysics from the University of Oxford and a DSc degree in physics from the University of York. He is a fellow of the Institute of Physics, the Royal Society of Chemistry, the Royal Society of Biology and the American Crystallographic Association. He was recently elected an honorary member of the British Biophysical Society.

In the end, it's extra effort that separates a winner from second place. But winning takes a lot more than that, too. It starts with complete command of the fundamentals. Then it takes desire, determination, discipline, and self-sacrifice. And finally, it takes a great deal of love, fairness and respect for your fellow man. Put all these together, and even if you don't win, how can you lose?

Jesse Owens (1913–1980)
Four-time Olympic gold medallist

Section I

Introduction: How Do You Know You Are Suited to Be a Scientist

You may be a scientist already engaged in research, development and discovery. You may be a person considering starting science as a career. You may be at school deciding between science subjects versus arts and humanities subjects. This book will describe for you what science is like, including the successes, as well as the challenges, and yes, some trials and tribulations. In your wider reading about science as a career, you will encounter books that describe how the skills that you will learn are transferable skills, and that no doubt is true. But I will not be describing how you would get along as a scientist-with-skills become banker or accountant or computer employee or detective. As a scientist, you will naturally wish to see your discoveries put to good use, and for other discoveries, you will have no conception as to their application, as they may well be too fundamental and basic science to be applied yet. Your work as a scientist in all its forms will be important, if you are good at selecting what to work on. You know, or simply sense, that your own skills are not to be neglected. You may also be sceptical that the softer skills beyond your laboratory bench are relevant; nevertheless, take a look; you may well be surprised. At some point, I will need to state a caveat and may as well do it here at the start; my book is a guide and cannot offer you rules or recipes.

The thought 'how do you know you are suited to be a scientist' could well arise at two stages of your scientific career:

1. You are thinking of becoming a scientist, which can occur when you choose an undergraduate BSc degree course (or BA, as my alma mater, York University, would describe my degree in physics) or an even bigger step,

whether to undertake a PhD (or DPhil, as my alma mater, Oxford University, would have it).

2. You have doubts; this can be of the 'I am not sure I am cut out for this', which may arise during stressful challenges, especially if you or your laboratory make a mistake, or more likely if you are frustrated by a career ceiling, that is, when you cannot secure promotion especially if you have tried several times.

Suffice to say, are you naturally curious? Or a better way of putting it, are you curious about nature? If you are then, yes, you are suited to becoming a scientist. This is a fundamental point! You then need to assess your evidence of being good at science or mathematics.

In school, in England, I faced the usual specialisation decisions at age 15 and then at age 18. At 15 years old, one has to decide between arts and humanities subjects versus science and mathematics subjects. But I straddled both! My slightly better subjects were history, geography and mathematics. A crucially timed parents evening, which I attended with my parents, led to a crucial decision; the careers adviser, who was especially on duty for the parents evening, advised us that there was a much better chance of a job in chemistry. We discussed this back home, and so I chose chemistry, mathematics and physics. My general studies were fortunately not abandoned, as the school policy was to also have pupils undertake a general studies curriculum, although I quickly found that a better way of learning this subject was to read a quality Sunday newspaper each week. The career adviser not only was commenting on the buoyancy of the British chemicals industry, which therefore provided good job opportunities, but also had a gleam in his eye that I might receive my school senior chemistry prize when I was 17 years old, which I did. At that age pupils then have to choose a university degree subject, and I chose physics, this time fascinated by the role of mathematics in its formulations and interpretations of the natural world. The University of York awarded me a first class honours degree in physics in 1974. My doctorate was in molecular biophysics, and the laboratory of my supervisor was based in the psychology/zoology building of Oxford University. Subsequently, I did hold faculty positions in the physics departments at Keele University and later back at York University, totalling seven years. But guess what, in 1989, I was promoted to a professorship, in chemistry, and served in that role until my (semi-)retirement in 2012. I suppose I can rationalise that I steadily worked against the forces of specialisation during the whole of my scientific life. How did I continue my strong interests in history and in geography you wonder? For history, I enjoy reading political autobiographies, and for geography, I enjoy travelling, which has included much of the whole globe. Academic training is necessary for a career in science, but it is not sufficient for you to know if you will really follow a career of scientific research and discovery. One of the crucial options for deciding that comes with the first job after the doctorate, namely the postdoctoral post. Not least, it will be your first salaried position, and your success in it will determine if you embark on a scientific research life's work.

A unifying aspect for any stage of our science interests and development, from school days through to retirement, is to reflect on what the scientific method is. The National Physical Laboratory based at Teddington, United Kingdom, has an explanatory card describing the scientific method, which I reproduce with their permission. They also have an extensive, helpful website on a wide range of scientific topics, for those coming to this book from any background and who are curious about this activity called *science*.

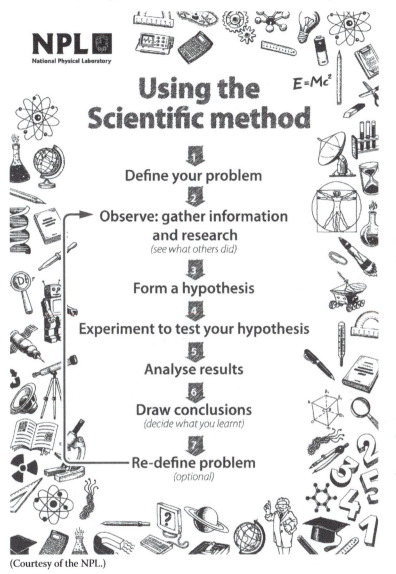

(Courtesy of the NPL.)

Figure: The National Physical Laboratory (NPL) based at Teddington, United Kingdom, has this lovely card explaining the scientific method. Notice of course the beautiful iconic scientific images decorating the words.

Overall, we can join the wonderful process of science and discovery. If we hone our skills to the best possible, we contribute as fully as we can to a scientific life.

Now, read on.

1 How to Undertake That Very Important Postdoctoral Research

After your PhD you may well consider taking your career in science further by look-ing for a postdoctoral research post. There are two types of this. The first is where you would apply to be a research assistant in response to an advertisement of some kind by a principal investigator (PI). The second is where you apply to an agency for funding for your own research proposal; with this option, you would, in parallel, be arranging for a research institution to base your fellowship at if it were funded. The first type is the predominant one by number.

With either type of postdoctoral post, there will be a set of research objectives. You will have to have the requisite skills essential and desirable, to do the work. These skills will closely link to the specific research skills that you learnt during your PhD. With that said, you will look to the new post to extend your skills and experience. Except in rare cases, you will, more or less, stay within your field of research. You would have more freedom for adventure and diversity within your own fellowship than within a research assistant post whose job objectives in any case are in the control of the PI and approved by the funding agency. There will most likely be specific deliverables expected of the research grant. If you are lucky enough to have won your own postdoctoral fellowship, then you will simultaneously be the worker and the PI/risk holder.

At the end of your first postdoctoral post, you will be considering very carefully either a second postdoctoral post or a tenure track post. The latter will have a four- or a five-year duration, whereas the postdoctoral job duration is typically three years, or even two years or one year. If you are in a second postdoctoral job, to consider a third such position is a very serious step, and indeed the perpetual postdoc is a career expression, meaning always being on soft funding agency money and therefore with no job security.

To obtain job security, and thereby with it your guarantee to follow your own research dreams, you need a permanent academic job or a job in industry. To reach that point, one can define an ideal career timeline of your PhD, your first postdoc-toral post (say two or three years long), your own fellowship (four or five years) and then a permanent post. This last step is arguably the most difficult, and well-recognised, scientific career transition. There are variations on this ideal which can still lead to a longer-term career in science.

How do you optimise these stages? A very often-quoted piece of advice is to move on: 'Do not stay in the same lab where you did your PhD'. Or 'do not stay in the same lab where you did your first postdoctoral post' or 'do not stay in the same country for each stage'. My own experiences were as follows:

- DPhil at Oxford University
- Postdoctoral post in the same lab in Oxford University
- A five-year post at Keele University
- Joined the UK's scientific civil service as a government employee on my first permanent scientific post
- Joined York University as a lecturer in physics on my first permanent academic job

All these were in the same country, the United Kingdom. So the generic advice that I started this chapter with, I did not follow in my own career! True, I did move universities, and I hopped between the university sector and the scientific civil service, both in and out. Possibly the most important job move with respect to this chapter was that I left the scientific civil service in late 1985, aged 32 years, to again take up a post as a lecturer at a university so as to have the right to submit research grant proposals, which were at that time denied to scientific civil servants.

This means that there are no fixed rules in making your career timeline. Suffice to say, it is important, though, for you to recognise and try and develop your own research vision and how to develop yourself and your skills. The governments and their funding agencies all regularly say that they want to encourage schoolchildren into science and that society needs science. But like so many things, saying and actually doing, both by governments and by their research agencies, are two different things. You will also need some luck then. I was lucky. I look back now and, being still young at heart, I think the chance for me to continue to contribute to science research and discovery is a wonderful thing, and so the world can really be one's oyster.

To support you in your work and endeavours, your employer will run courses and skills training. These will build on your researcher development training within your PhD; for an example, see what the University of Manchester offers a prospective research student [1]:

> Each of our four Faculties offers an extensive researcher development programme, providing a blend of cross-disciplinary and generic skills training opportunities to equip you with the skills, attributes and knowledge to thrive as independent researchers and professionals. Training covers areas such as research management, personal effectiveness, career management, engagement, influence and impact.

And from Dame Professor Nancy Rothwell, president and vice-chancellor, the University of Manchester:

> Skills training has always been important for researchers and never more so than today. While researchers must have knowledge and skills in their own research area, they must also master the critical skills of communications (via multiple routes and to multiple audiences), team-working, leadership, critical thinking and much more. All of these are essential for a successful researcher but also for the many other career paths that researchers may choose to take and for success in life in general.

You should build on your skills training that you will hopefully have gained in your PhD. Firstly, set aside some time each year to take advantage of these during your postdoctoral research. Secondly, you can sign up for the mentoring schemes again that your employer will run; I discuss the benefits of mentoring in Chapter 24. Thirdly, you can join the professional society which supports your subject discipline; for me, that was the Institute of Physics, the Royal Society of Chemistry and the Biochemical Society. Late in my career, I also joined the Royal Society of Biology. Fourthly, the organisation Vitae (https://www.vitae.ac.uk/) offers many useful pamphlets and training events. So, good luck.

REFERENCE

1. The University of Manchester (n.d.) Postgraduate researcher development. Accessed 7 August 2016. Retrieved from http://www.manchester.ac.uk/researcher-development.

Section II

Skills for a Better Researcher

2 How to Recognise a Good Idea

The first step to recognise a good idea is to have an idea. I wonder how Newton, when the apple fell from the apple tree, said to himself, aha, gravity, and let us write down an equation for the attraction between two heavenly bodies. Or I wonder how Einstein suddenly thought, what if the speed of light was not infinite? To encourage ideas to arise is a skill in itself. I found out that I had many ideas while cycling or hoovering the carpet. Ironing was hopeless as the semidreamlike state required led to burnt shirts. A semidreamlike state while cycling was also to be treated cautiously, but the increasing number of cycle tracks without cars or sometimes even pedestrians have surely proved to be a boon to creativity! Probably my biggest idea was not had in an instant but grew during the first year of my DPhil, namely that the lab equipment and the physics procedures where I was were way too primitive. So, my *idea* was that something is wrong here. I shared this with my supervisor, Dr Margaret Adams, who I am glad here to pay tribute to her skills as a supervisor, for she took it all very well and as part of her careful response arranged for me to meet her supervisor, the Nobel Prize winner (Chemistry 1967) Professor Dorothy Hodgkin. Dorothy reminded me of the basics; that the field of crystallography involved the three physics probes of electrons, neutrons and X-rays. So far so obvious, but it is always good to come back to the basics! Dorothy then did the most generous thing and shared with me a letter she had received from another senior scientist, Sir Sam Edwards, who in turn had received a letter from Sir Ron Mason about new work at the synchrotron in Stanford, California. This gave a glimpse of the world for X-ray crystallography that I was groping towards. Sir Ron's letter described other senior scientists' views of the matter as controversial. There was no controversy in my mind that this was a *good*, indeed, *great* idea. Pretty much everyone I met did not believe this. In fact, at my DPhil interview, one of my interviewers actually stated that there was no point in my joining this research field as a physicist as the existing methods were just fine. I do not really know why I was not put off, but rather regarded it as some sort of odd, eccentric person I had met. The synchrotron became the majority of my science research career!

To decide if an idea is good is a critical step. Einstein's life is beautifully documented in Abraham Pais's biography of Einstein [1]; it has the uncommon style as biographies go of interleaving Einstein's life with chapters on his publications and discoveries. A significant feature of Einstein's 1905 special relativity paper [2] is that it has no cited references. Pais elaborates at length about this, in particular, discussing the correspondence that Einstein had with Lorentz and Poincaré. Late in his life, Pais documents, Einstein acknowledged that Lorentz did stimulate his discovery of special relativity. With all this said, Einstein's Nobel Prize for Physics of 1921 was not for his theory of special relativity, for which the award presentation by

Professor S. Arrhenius, chairperson of the Nobel Committee for Physics of the Royal Swedish Academy of Sciences, on 10 December 1922, stated:

> There is probably no physicist living today whose name has become so widely known as that of Albert Einstein. Most discussion centres on his theory of relativity. This pertains essentially to epistemology and has therefore been the subject of lively debate in philosophical circles. It will be no secret that the famous philosopher Bergson in Paris has challenged this theory, while other philosophers have acclaimed it wholeheartedly. The theory in question also has astrophysical implications which are being rigorously examined at the present time.

Instead, the conclusion of the award presentation was that the Nobel Prize was for his theory of the photoelectric effect, also published in 1905.

> Einstein's law of the photo-electrical effect has been extremely rigorously tested by the American Millikan and his pupils and passed the test brilliantly. Owing to these studies by Einstein the quantum theory has been perfected to a high degree and an extensive literature grew up in this field whereby the extraordinary value of this theory was proved. Einstein's law has become the basis of quantitative photo-chemistry in the same way as Faraday's law is the basis of electro-chemistry.

Was Einstein's paper on special relativity a good idea? Well, obviously yes, it has had a huge impact. But could he have explained its potential impact then to a research funding agency? That is, as is now required with a research grant proposal. Einstein, one could argue, needed only a pencil and paper, and it is not a real example for today. Well, what if he had no salary, had no means of support and needed a research fellowship to undertake it? This example shows up a number of things. Firstly, it shows how difficult it is to decide if an idea is good; one has to define *good*. Secondly, we might decide that an idea has to have low research risk, and if something goes wrong, what are we to plan to do instead? Indeed, it is argued that research agency funding committees are risk averse, and fortunately, this has encouraged funding earmarked for adventurous research (see Chapter 4).

What if you find that your idea has already been had by someone else and partly or fully taken to a research conclusion? This is an easier question. Firstly, of course, it helps you realise that this idea was most likely good after all. In the research undertaken by the other person or research group, there may be shortcomings in their approach or the field of activity is so large that there is plenty of room for your research group as well. It is very important to carefully, and as generously as possible, refer to the other person's work. After all, you may well have hesitated to commit any resources to it in the first place. Of course, you may simply have been scooped yourself! The answer to this problem is to have plenty of ideas. An example of someone with many ideas was Nikola Tesla, the scientist whose name is now honoured as the unit of magnetism; for example he has a very large number of patents, over 300 (see https://en.wikipedia.org/wiki/List_of_Nikola_Tesla_patents).

I have explained how it is not so easy to decide early on if an idea is a good one. The moral of the earlier stories is that you as the researcher must trust your own judgement. But let us look at this with some modern management tools. Ideas lead to some objectives in a plan of action. What are good objectives? Management tools and planning state that these should be *smart* objectives, namely

- Specific – Target a specific area for improvement.
- Measurable – Quantify or at least suggest an indicator of progress.
- Assignable – Specify who will do it.
- Realistic – State what results can realistically be achieved, given available resources.
- Time-related – Specify when the result(s) can be achieved.

These are helpful criteria to assist you to evaluate whether an idea is *smart*. One common objection to an idea is *ahead of its time*. This is the time-related aspect. As a more detailed example of my own, I select one when I was pioneering one of the new methods and instruments for X-ray crystallography at the UK's Synchrotron Radiation Source in the early 1980s. I investigated the use of much longer X-ray wavelengths than was usual. My tests worked well, but I had no immediate challenge I needed to solve with them. Much later, nearly 20 years later, in the late 1990s, I had the perfect challenge to apply the new methods and instrument to. This led firstly to the publication of a new protein crystal structure [3] and subsequently more details of my tests done 20 years earlier [4]. This work was featured in the citations of my prizes that I received from the American and European Crystallographic Associations in 2014 and 2015. So this means that you have to adopt a generous view of what is time-related about when the results can finally be achieved. This study would not have sat well in a research grant proposal, which would expect clear milestones and a clear conclusion after say three, maybe five years of funding, but at that time, this work sat well within my role as a scientific civil servant where I did the foundation research and development myself.

These two types of ideas, one where the idea and the early work led to the application only after many years, and another where dissatisfaction encourages progress, are not necessarily the most common. The more common is likely to be where you know the idea is good and the first funding agency you try to obtain funds to tackle it also judges it as internationally excellent, but there are insufficient funds available and so you face the ranking order problem. There are simply higher-priority proposals, judged to be of more significance, in the eyes of the funding agency, its committee and its referees. This naturally leads on to the next chapter.

REFERENCES

1. A. Pais (1982) *Subtle Is the Lord: The Science and the Life of Albert Einstein.* Oxford University Press, Oxford.
2. A. Einstein (1905) *Zur Elektrodynamik bewegter Körper* [On the electrodynamics of moving bodies]. *Annalen der Physik* 17: 891.
3. M. Cianci, P. J. Rizkallah, A. Olczak, J. Raftery, N. E. Chayen, P. F. Zagalsky and J. R. Helliwell (2001) Structure of apocrustacyanin A1 using softer X-rays. *Acta Crystallographica Section D-Biological Crystallography* D57: 1219–1229.
4. J. R. Helliwell (2004) Overview and new developments in softer X-ray ($2\text{Å} < \lambda < 5\text{Å}$) protein crystallography. *Journal of Synchrotron Radiation* 11: 1–3.

3 How to Make Significant Discoveries

Every great advance in science has issued from a new audacity of imagination.

John Dewey (1859–1952)
American philosopher, psychologist and educational reformer [1]

A clear way of judging if an idea can lead to a significant discovery is to ask yourself whether anyone would care about the results that you might obtain, beyond your own satisfaction. This is referred to in grant proposals as *who are the likely beneficiaries?* This criterion can be good and sits easily with research questions such as how to arrive at noncarbon energy sources or a cure for a disease. When the proposal is controversial, such as nuclear energy, how good an idea is can be very different depending who you talk to! France has always stuck with the nuclear energy future, the UK abandoned and recently readopted nuclear energy and Germany has recently abandoned nuclear energy for its future energy supplies. This seems to be, and is, a postcode lottery where the answer or the service you get varies according to the place and there is therefore no absolute answer. This is a useful reminder that some questions either have no easy answer or indeed cannot be answered.

A rather unsettling feeling I find as a scientist is when a politician states that he or she must take a judgement on insufficient data, which is anathema to a scientist. This is our strength, but it is also a weakness. One should be aware of the saying by Voltaire that 'the great is the enemy of the good', otherwise referred to as *perfection is the enemy of progress*. The skill of good scientific judgement, and whether an idea is good or not, is whether it would involve acquiring detailed results that are too minute and pernickety to be of any real value. A balance has to be struck between dotting i's and crossing t's type of data versus having insufficient data to make any clear conclusion or a significant discovery.

Having embarked on a research project, you will have your results and want to write them up. This will lead to the now usual discussion among your authors as to which journal to submit to and for which one criterion will be the journals' impact factors. Some journals such as the *Public Library of Science* (PLOS) journal is interested only if the research was done properly and is described properly, that is with no descriptions of the significance of the results. The referees commissioned by most journals, except *PLOS*, will ask the referees to score the likely impact of the article. This will be assessable finally after publication by the number of citations and article downloads. Curiously, in my experience as an editor, there is a surprising lack of correlation between these two metrics. A formal study of this, for over 100,000 physics articles [2], shows that there is indeed a weak correlation, at 40%, and with a mean ratio of downloads to citations of 2.24. However, Brody et al. (2006) [2] also note that 'few articles with high citation impact receive low downloads impact'. They also believe that once the proportion of open access articles increases from the then 20%, the correlation between downloads and citations will likely increase and that the correlation will no doubt vary from field to field.

Let us look at a classical definition of *significance*, for example, that is described by Kuhn in his book, Chapter XII, page 144 [3]:

> What is the process by which a new candidate for paradigm replaces its predecessor? Any new interpretation of nature, whether a discovery or a theory (or a method), emerges first in the mind of one or a few individuals. It is they who first learn to see the science and the world differently, and their ability to make the first transition is facilitated by two circumstances that are not common to most other members of the profession. Invariably their attention has been intensely concentrated upon the crisis-provoking problems; usually, in addition, they are men (and women) so young or so new to the crisis-ridden field that practice has committed them less deeply than most of their contemporaries to the world view and rules determined by the old paradigm. How are they able, what must they do, to convert the entire profession or the relevant professional subgroup to their way of seeing science and the world? What causes the group to abandon one tradition of normal research in favour of another? Therefore, paradigm-testing occurs only after persistent failure to solve a noteworthy puzzle has given rise to crisis.

Therefore, a good way to see a significant chance for change is where there is a puzzle or even a controversy or a crisis in a field. In my own field of protein crystallography, I made a few such observations of significant chance for change and which have consistently been controversial at the beginning at least. These developments seem to me to have done my career, and those of various colleagues around the world, a power of good by diving into those areas.

Another important measure of whether a scientific research advance is significant is whether it has impact. This can be very obvious in retrospect. Synchrotron X-ray radiation has had a huge impact on the great expansion in number, complexity and quality of the crystal structures of biological macromolecules these last 40 years; this is readily measured via depositions in the Protein Data Bank and the synchrotron beamlines from which they were derived. Planning for impact is increasingly a wish of the research funding agencies. A handbook to assist in this is available [4]. In retrospect, I clearly made a good impact plan as defined by Mark Reed [4] for my synchrotron research and development for protein crystallography to have maximum impact; the core of an impact plan is *knowledge exchange* with relevant stakeholders. So, as well as my instrumentation and methods research and development through the 1980s, I undertook several other roles: I wrote a textbook on the subject [5], served as the leader of the European working group for this field at the planned European synchrotron and on various other such advisory roles in the United States and Japan, as well as engaging with the pharmaceutical industry from the outset on such developments (e.g. see my article in Ref. 6).

REFERENCES

1. J. Dewey (1930) The Copernican Revolution. In *The Quest for Certainty: A Study of the Relation of Knowledge and Action.* George Allen & Unwin Ltd, London.

2. T. Brody, S. Harnad and L. Carr (2006) Earlier Web usage statistics as predictors of later citation impact. *Journal of the American Society for Information Science and Technology* 57(8): 1060–1072.

3. T. S. Kuhn (1996) *The Structure of Scientific Revolutions*, third edition. The University of Chicago Press, Chicago.

4. M. Reed (2016) *The Research Impact Handbook*. Fast Track Impact, Huntly, Aberdeenshire.

5. J. R. Helliwell (1992) *Macromolecular Crystallography with Synchrotron Radiation* (paperback: 2005). Cambridge University Press, Cambridge.

6. J. R. Helliwell (1983) Protein crystallographic drug design using synchrotron X-radiation. *Acta Radiologica* Suppl. 365: 35–37.

4 How to Write a Successful Grant Proposal

This chapter you might decide should have come before the previous chapter simply on the assumption that you always need research funds to undertake a piece of research. I maintain that this is definitely not the case, and I gave examples in Chapter 2 of how one can proceed with an idea, namely you do the work yourself, as I did in my examples. In the more business-oriented, current university research environment, it seems almost frowned upon to undertake any research that is not funded, and in turn, although it might be funded, it might not be from a funding agency that will pay research overheads. There is then a pecking order as far as the university administration is concerned: funding agencies that pay overheads are better than those agencies that do not and, in turn, are better, much better, than doing the work yourself. At the University of Manchester, one of the most famous recent examples of doing the research yourself was the discovery of graphene, which resulted in the Nobel Prize for Physics to Professor Andre Geim and Dr Konstantin Novoselov. They relate the story of how they would reserve Friday evenings for adventuresome science, and one such example was peeling a monolayer of graphene by using sticky tape from a graphite pencil! I cannot offer such an exciting personal story. That is, no Nobel Prize just yet, but likewise, I would pursue adventure research at the synchrotron during the single-bunch, low-current operating mode, which was beamtime of no interest to my outside-university users. Those days could have been my days off. But instead, I would also work those days at the synchrotron as well.

But a large portion of your time as an academic will be devoted to preparing research grant proposals, on your own or collaboratively.

There are many different funding agencies. You will need to study what each wants from the research it funds before you consider submitting an application. Obviously, you will also learn the fine details of the structure and the layout of their proposals. There is little point in your trying to impose your will or views. Secondly, research-active institutions these days should provide a means for you to receive detailed comments from experienced colleagues on your draft proposals. Thirdly, you should not prepare a grant proposal in haste; I will go into the details of time management skills in a later chapter, but suffice to say, a grant proposal is very important and should not be undertaken as if it were urgent. You should be in full control of this and therefore not rush it. Fourthly, it may be the case that you must show and demonstrate that the necessary feasibility experiments are mature; in essence, all agencies these days basically wish to minimise the risk of a proposal not proceeding as expected, following the milestones that you will have to describe. This latter point is contentious in that it leads to funding agencies being accused of not supporting the truly innovative research. The Wellcome Trust had a scheme that ran for a while for adventurous research proposals. This was obviously highly oversubscribed, clearly fulfilling a need! With a colleague, I received one of these awards; the proposal success rate was only 1 in 30!

I only ever made one complaint to a funding agency research manager about the handling of a research grant proposal I had made. There were six referees' reports, all except one were very favourable, which stated, 'I don't care how the protein crystal structure is going to be solved, and too much of the proposal is devoted to that, indicating that there is a worry that it cannot be carried out'. In effect, the implicit risk was deemed unacceptable. The rejection letter quoted only this one sentence. How should I then proceed? Well, I was very encouraged by the five other reports which added to my own judgement to press ahead. By coincidence, I was fortunate to receive a sparklingly good application from someone for a PhD place with me; I replied within half an hour of my receiving the e-mail, the applicant recalled later. I outlined an interesting project; all my details were to hand, and with such a good applicant, I secured a PhD studentship for him. Since the applicant was from the European Union (EU), the funding agency's specific rules for EU applicants meant that this covered only the university's fees. For the subsistence, I secured a faculty contribution (two-thirds) and a further department contribution of one-sixth, and I found the final one-sixth. At this stage, I had no monies for consumables! They had to be found later. This research led to a publication in the *Proceedings of the National Academy of Sciences* (*PNAS*) [1], which was highly downloaded in the first six months after publication (654 times). Indeed, I received a letter from the *PNAS* Editorial Office thanking me for submitting this research to them. The results were widely quoted in the national newspapers, as well as on radio, and later were written up by specialist science writers in *Physics Today* and *Swiss-Prot*. The moral of this story was that I and the PhD student ourselves took on the risk that the funding agency would not.

Overall, key points to remember in how to write a successful grant proposal were nicely summarised by my former PhD student, Dr Titus Boggon (now at Yale University), when he read the draft of this chapter, which I quote now with his permission:

- Careful reading of a research programme announcement can be important.
- Follow the grant guidelines.
- Sell the proposal – the grant reviewers have a large stack of grants, and they are busy, so a well-written and nicely formatted grant will help put reviewers in a positive frame of mind.
- Pretty figures to help make the text more digestible are a good approach.
- For early career scientists, the experience of serving on review panels will help your own research grant proposal writing – so try to get on review panels as fast as possible.

REFERENCE

1. M. Cianci, P. J. Rizkallah, A. Olczak, J. Raftery, N. E. Chayen, P. F. Zagalsky and J. R. Helliwell (2002) The molecular basis of the coloration mechanism in lobster shell: β-Crustacyanin at 3.2 A resolution. *Proceedings of the National Academy of Sciences of the United States of America* 99: 9795–9800.

5 How to Assess Research Risks

In the previous chapter, I introduced the term *risk*. I have found the quantification of risk a particularly powerful tool to guide one's thinking, but note, not to decide one's thinking. Firstly, though, we have to distinguish risk assessment as a management tool, as distinct from areas of chemical safety, or chemical hazards, or biological safety and hazards. The laboratory safety laws are outside the scope of this book, but suffice to say, the local rules of your institution and the laws of the land in the country where you live must be adhered to.

I first had formal management training in risk assessment as a management tool when I became director of the UK's Synchrotron Radiation Source. When I returned to the University of Manchester, I undertook further voluntary training in risk assessment. Basically, one proceeds as follows.

In any activity such as a research project plan, there will be risks of some sort which you can analyse in a systematic way, which is known as *risk analysis*. Indeed, there will most likely be more than one risk, and you can quantify each one and place them into a ranking order of seriousness in a folder of risks for that project. For each risk, you proceed as follows:

- You identify what its impact would be on your project if it actually happened, and you score this on a scale of 1–10, with 10 being the most serious.
- You then assess how likely it is to occur, its probability, and you score this on a scale of 0–1 (0 means it will never happen and 1 means it is certain to happen).
- The product of these two numbers is then your numerical estimate of the risk.
- You need a plan to deal with a risk if it were to occur; this final step is sometimes referred to as the *risk mitigation plan* for any given risk.
- You need to identify a single person who is responsible for a given risk and the risk management plan; this person is the *risk holder.*

I illustrate a risk's estimation in a simple graphic in Figure 5.1.

In diligently managing risks in a project, it is important that you and those involved with a project to

- Agree the folder detailing all the identifiable risks.
- Review the risks folder for this project at least annually.
- If necessary, introduce new risks and/or delete expired risks as a project proceeds.
- Reaffirm the risk holders in this annual process.

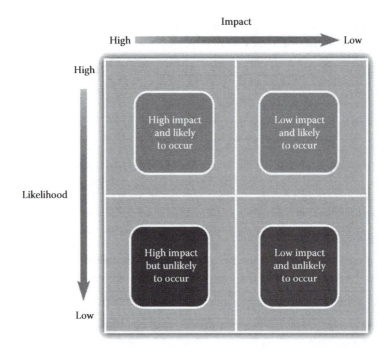

FIGURE 5.1   A simple graphic to illustrate how to view a given risk. The worst risks are of high impact.

The estimates of probability and of impact scores for a given risk are necessarily subjective, but you may well have a fairly good idea which risks are higher than others, and so you can adjust the numerical values accordingly to at least make a realistic ranking order.

As an example, let us analyse in general terms a science journal. This is something familiar to all of us whatever scientific field we are in. (If I were to select a given research project and analyse its risks, it would likely get too technical in the details to be of use here as a case study.)

There is a risk that

1. If the accuracy of the science content of a journal is not sustained, then the reputation of the journal would be damaged: probability: 0.7, impact: 10, overall risk: 7. Risk holder: the publisher. Mitigation strategies: Regularly consult a journal's readers and authors via questionnaires to monitor their levels of satisfaction. Also ensure that proper procedures for the selection of expert referees by the editors are being followed.
2. If the accuracy of the scientific data underpinning the science content of a journal is not sustained, then the reputation of the journal would be damaged: probability: 0.7, impact: 10, overall risk: 7. Risk holder: the publisher. Mitigation strategies: Regularly consult a journal's readers and authors via questionnaires to monitor their levels of satisfaction with the scientific data

linked to the article. Also ensure that proper procedures for the selection of expert referees of the data by the editors are being followed. It would be expected that the referees of the data would also be the referees of the submitted article.

3. Poor production quality of the published articles would lead to readers, authors and subscribers being discontented, leading to financial losses: probability, 0.7, impact: 10, overall risk: 7. Risk holder: the publisher. Mitigation strategy: Regularly consult a journal's readers, authors and subscribers via questionnaires to monitor their levels of satisfaction.

4. The marketing, or the web presence and the availability of the journal titles, is inadequate, thus leading to a reduction in the number of articles submitted to the journal, destabilising the readership and the subscriber base: probability: 0.2, impact: 7, overall risk: 1.4. Risk holder: the publisher. Mitigation strategy: Promote the journal actively including at least having not only promotional leaflets and cards at the main international conferences but, ideally, also a manned stand in the commercial exhibition at these conferences.

5. If the journal does not adapt to change its science scope as a field develops, then the journal would look out of date and will possibly develop a poor impact factor: probability: 0.2, impact: 7, overall risk: 1.4. Risk holder: the editor-in-chief of the journal. Mitigation strategies: The editor-in-chief should attend relevant international conferences. Also, the editor-in-chief and the publisher should work together to refresh and clarify the scope of the journal.

I have placed those individual risks for a journal in a numerical ranking order. They would comprise the risk folder for the journal and create priorities for action.

As an example of a research journal which I have found to be an exemplar of good practice with respect to refereeing of submitted articles and their accompanying diffraction data and molecular coordinates embedded in a crystallographic information file, I would mention *Acta Crystallographica Section C Crystal Structure Communications* (renamed in 2015 as *Acta Crystallographica Section C Structural Chemistry*). My wife, Dr Madeleine Helliwell, was a coeditor for several years for this journal, and so I saw at first hand just how hard she worked with each article, data and atomic coordinates, as did her selected referees, to ensure that everything was as perfect as possible. Risks 1 and 2 were minimised!

Generally, in project management practice, the overall situation on any risk is judged as 'green' for continue with project, 'amber' for review urgently and 'red' for stop the project. A weakness of the risk analysis method is that it is not easy to quantify the effect of the coincidental simultaneous occurrences of the identified individual risks. It is also worth stressing that the cost of a mitigation plan of action should not of course exceed the cost of a given risk coming to fruition! One should also not become so obsessed with the risks that a good project never starts; after all, that would be akin to never crossing the road for fear of a car or a bus hitting you. However, with your risk analysis tool, you now know your mitigation plan to deal with that particular risk; you would cross the road only at the pedestrian light-controlled crossing traffic lights!

Let us make a general case study example to complete this chapter. Your project is your plan of how you will travel to work. You prefer to commute by bicycle, as it is a way of helping you keep fit and might well be about as quick as by car or public transport. But you are naturally concerned to plan your cycling route carefully in terms of your overall safety on the road. These days, with the government's making data available in an open way (open data), you can find the road accident statistics for cyclists in your area; my hometown is Stockport, and I live 4 1/2 miles from the University of Manchester. The data for cyclist road accident statistics in Stockport, which would cover part of my route if I used roads, are available on the web [1]. You also find out that there is a route for cyclists without any traffic, for a good part of the way [2]. So you obviously select the route with a traffic free portion and the roads with the lowest probabilities of road accidents to connect to the start and the end of the cycle route. Another interesting aspect about travelling back from work concerns how tired you are, which is linked in turn with your hours of working. Simply put, one should not cycle or drive when overly tired.

REFERENCES

1. Greater Manchester Transportation Unit (GMTU) (2008) Stockport child and adult cycle injury accident statistics. Report number 1394. Accessed 7 August 2016. Retrieved from http://www.gmtu.gov.uk/reports/gmtu_report_1394_transport_statistics_stockport _2007.pdf.
2. SUSTRANS 'Fallowfield loop' cycle way map. Accessed 7 August 2016. Retrieved from http://www.sustrans.org.uk/ncn/map/route/fallowfield-loopline.

6 How to Set Up, Lead and Care for Your Research Team

The major occasion when you have the chance to establish your laboratory (lab) is when a university, most likely, entrusts you with establishing its presence in an academic discipline, and, with it, they will appoint you to a professorial chair of a subject. In my case, it was at the University of Manchester, and, in a very competitive selection process, I became the university's professor of structural chemistry. I held this post for 23 years, commencing in January 1989 and continuing until August 2012. I was quite a young age of 34 when I was offered the post. With your offer, as with mine, there will be monies for your start-up equipment and associated posts, academic and technical. The starting monies, most likely, will be insufficient to equip your opening phase of research work completely, and so you will need to make research grant applications to funding agencies for the additional equipment, and running costs and consumables. The recruitment of staff further adds to the administrative load, but a general feeling of excitement will carry you through these early years; yes, years not months! You will also have time on your hands, and I used mine to write a research monograph book, which did well in becoming a paperback edition eventually. You will also find yourself consulted on many new matters, both external and internal to the university.

Leading your team will likely involve matters of both arriving at consensus and following your own judgement, perhaps without consensus; this I think is inevitable. Either way, communication with your team and with individual team members is vital. A meeting as a group, as a simple option, includes a journals meeting. My version of this was to meet every month having asked each student to select a recent publication and circulate a copy to each member of the group to read beforehand as well, and at the meeting, the student firstly describes it, critically as may well be required, and then opens the paper for discussion. This provides a dynamic to the group as a whole and helps your lab members develop critical skills of communications, as well as critical thinking, mentioned by Professor Nancy Rothwell who I quoted in Chapter 1.

PhD and masters student progress meetings work best with individuals as a one-on-one meeting, but some projects obviously are team efforts, and joint progress discussions are also a must. You are the chair of these meetings, or if it is a collaboration with another lab, it will be jointly chaired. At the end of the meeting, you must restate the agreed points and actions. With the advent of digital camera smartphones, I would write these agreed points and actions on a white board, take a photo and e-mail to all those present at the meeting and, with their agreement, to others who also needed to know the meeting summary. I will discuss chairing meetings again in Chapter 21, but mainly from the perspective of advisory board committees etc., rather than your own lab meetings.

In caring for your team, do not micromanage it. You have appointed them because they are good at what they do. Your PhD students will offer a much greater variety of challenge to you as their educator. In the UK system, the PhD typically lasts for three years, quite short compared with continental Europe. This is changing with funders now supporting PhD students for 3 1/2 years and even 4 years. A very good PhD student will need full supervision in the first year, a mixture of supervision and freedom to follow their own judgement and ideas in their second year and in their final year you can trust them to ask you for guidance rather than you carefully checking each detail. There will always be the need for regular discussions, at least weekly. The very good students will most likely, but not always, be seeking to continue in research as a career. My past students include ones who are now professorial rank and/or senior researchers in research institutes or research-driven organisations. The medium- and lower-rank performing PhD students will obviously require you to vary your pattern of supervision that I described earlier for a very good PhD student. Basically, you may find that you need to very much guide such PhD students in their second year, and even into their third year. Over the years universities in general and the University of Manchester in particular have become evaluated on their PhD submission rates, namely the percentage submitting their PhD thesis before a four-year deadline since the PhD student started. This has meant that the PhD student progression from first to second and second to third years has become more formalised. Thus, a written transfer report and an oral examination of it as well by independent academics, internal to your department, are needed. These anyway have been a good initiative in my view, as both the report and the oral examination of it are good practice for the PhD student for their PhD thesis writing and the PhD oral examination itself. Your difficult cases will be those who are very weak at the examination. These will undoubtedly have needed a lot of help from you both through the previous year and with their written report. The most difficult will be those who fail the end of the second year examination. Such cases should have been weeded out at the end of the first year, but this might not have been the case for a variety of reasons. Clearly, the time and the financial commitment at that stage of both the student and yourself after two years will be a fairly severe penalty to give up on but can in fact be the best thing to do for both sides. I use the word *sides* as that is what it will most likely have become by that stage. For this reason, another excellent development is that a PhD student has not only yourself as supervisor but will also have an advisor independent of you and your research lab. You will also of course be an advisor to other PhD students who are not your own, which will give you the necessary full range of perspectives. I further deal with this topic in Chapter 23.

Over the years, as you become known to the faculty and then the central university administration, you will no doubt be called upon to serve on student disciplinary committees. I served on many, mainly to decide on student plagiarism or cheating at undergraduate and master levels. But one case was a PhD thesis involving plagiarism of another PhD student's work; one research chapter was unplagiarised. The case was complicated, however, by the fact that the supervisor had abandoned the PhD student, and the department had appointed the advisor to be the supervisor instead. Now the crucial point is that no replacement advisor to the PhD student was appointed. The disciplinary committee was faced with a very awkward choice, to refuse the student to progress and be expelled, or because the student had not been properly advised, the student had to be given the option to revise his/her PhD thesis, substantially obviously. Naturally, we did not learn of the outcome, although most likely, it was that the student could only submit for a master's degree.

The skills you need as a lab leader in general and as a PhD supervisor in particular would largely be learned on the job. Gradually, this has changed, and training has become formalised, with compulsory and voluntary training elements. Thus, when I was appointed a lecturer in 1979, I had an afternoon of training. When I was appointed a professor in 1989, I had two days of training, and an extensive handbook of guidance to supervisors was issued with yearly updates. When I was in the final phase of my career as a professor of many years of experience, I was appointed senior mentor for new academics. I am proud of this role that I was entrusted with, and I took it very seriously serving as senior mentor to more than 10 new academics. Each new academic undertook a year of detailed training with numerous course modules to attend, two lectures to be given with formal assessments and an extensive reflective portfolio report to be written, assessed by myself and a faculty officer. The UK government has recently announced that it will introduce a teaching excellence framework (TEF), a further formalised change in academic training. I confess to feelings of nostalgia and that even in the worst case of teaching undergraduate students poorly (not me, of course), was it such a bad thing that students would have to find out for themselves? A prominent example of such a case was commented on by Sir Peter Day in his autobiography [1] and that which I wrote a book review of [2]. He pointed out the poor quality of the undergraduate lectures in chemistry he received from Dorothy Hodgkin, Nobel Prize winner in Chemistry 1964 [3]! The prospect of the UK TEF failing Dorothy Hodgkin's teaching and thereby depriving those undergraduates of direct contact with such a frontline researcher runs against the basic principle of a university that teaching and research are, or should be, inextricably linked. This book is about skills you will note, including teaching, and so the modern university will have to embrace better teaching skills and yet including the not-so-good-at-teaching researcher. I further deal with this topic in Chapter 22.

As an extract of my book review [2] on this, my remarks are very pertinent to my still championing the importance of skills:

Of his days as an Oxford University undergraduate chemist Peter Day pulls no punches. First in his firing line [p. 36]:
    Several of the worst lectures I ever wasted my time on in Oxford were presented by people who made their own distinguished and lasting contributions to knowledge. Yet so wrapped up were they in the fine detail of their expertise that they were quite unable to imagine themselves in the situation of someone who wanted, even yearned, to have that door opened but needed help in lifting the latch. The time has come to name names: first up a Nobel Laureate. Dorothy Hodgkin was billed to induct us into the field which she had very much made her own, the crystallography of large molecules. This is of course a highly technical matter but profoundly significant...Whilst modest and charming, though steely of purpose in her working life, Dorothy ... took a few

slides from the box containing her research presentations and proceeded to share them. ... (thus) student attendance dwindled from a hundred to half a dozen.

Since Peter Day firstly knew that six students were left and also since he stated that he 'wasted his time with some lectures', we can deduce he was one of the remaining six! At this point I feel the need to defend, if not Dorothy's undergraduate teaching, her very positive influence on postgraduates as exemplified by her help to me when I was a D.Phil. postgraduate student with Margaret Adams in Oxford (1974–1977). In turn, Margaret had been Dorothy's D.Phil. student.

Continuing on Peter Day's comments on Dorothy's undergraduate lectures that he received, and his five fellow stalwarts who stuck it to the end, I do not doubt that as students they had excellent chemistry, maths and physics school exam scores, and so they would be my ideally trained and prepared undergraduates, and so Dorothy was obviously at fault in only focusing on research frontier material. But I would also presume to say that the Oxford University of the time was at fault as well in not providing adequate lecturer training or formalized lecture course monitoring. I would note finally that I also included research frontier material where I could, especially of the most recent structural crystallography results. In educational terms, as I see it, the aim and purpose of a leading university is to blend the best in core material with research frontiers' material. One without the other is unsatisfactory, as Peter Day relates. So, page 36 of Peter Day's book stimulated much thought in my mind!

Formalised training then is important. Let us look in more detail at the training for new academics at the University of Manchester [4]. I regard this as a 'Rolls Royce' level of training. In year 1, this involves course modules such as how to prepare a course descriptor, observation of your teaching by an experienced lecturer, how to use e-learning tools provided to students, conducting tutorials, supervision of postgraduates, securing grants and managing research projects, communication of results and understanding possible intellectual property possibilities. Overall, you would learn to develop your skills of critical reflection on all aspects of the academic role. In year 2, you would have a second teaching observation by an experienced lecturer, and you would have a day of training in specific aspects of teaching in your science discipline and start to think about writing the portfolio overview of your understanding of your roles. In year 3, the writing of your portfolio takes place in earnest and is assessed by the senior mentor.

The new academics programme handbooks for the Faculty of Medical and Human Sciences [5] and Humanities [6] in the University of Manchester naturally have the core elements regarding the roles of educator and researcher. With that said, there are major differences in practice between them, not least the treatment of people for the former, that is, medicine, and for the latter, no lab work.

A wide ranging description of leading your own laboratory is the book by Kathy Barker [7].

In summary, as a lab leader, you will need to fully embrace research and teaching if you are to fulfil at least my definition of a university.

REFERENCES

1. P. Day (2012) *On the Cucumber Tree: Scenes from the Life of an Itinerant Jobbing Scientist*. Grimsay Press, Glasgow.

2. J. R. Helliwell (2014) Book review of *On the Cucumber Tree: Scenes from the Life of an Itinerant Jobbing Scientist*. *Crystallography Reviews* 20: 2, 157–159.
3. D. Hodgkin (1964) Nobel Prize in Chemistry 1964 *'for her determinations by X-ray techniques of the structures of important biochemical substances'*. Retrieved from http://www.nobelprize.org/nobel_prizes/chemistry/laureates/1964/.
4. The University of Manchester (2010) New academics training programme for the Faculty of Engineering and Physical Sciences. Accessed 7 August 2016. Retrieved from http://www.academicsupport.eps.manchester.ac.uk/new-academics-programme.pdf.
5. The University of Manchester (2013) New academics training programme for the Faculty of Medical and Human Sciences. Accessed 7 August 2016. Retrieved from http://blogs.mhs.manchester.ac.uk/nap/files/2014/10/NAPHandbook2013-14.pdf.
6. The University of Manchester (2014) New academics training programme for the Faculty of Humanities. August 7, 2016. Accessed 7 August 2016. Retrieved from http://www.humanities.manchester.ac.uk/tandl/documents/HNAPHandbook2014-15_002.pdf.
7. K. Barker (2010) *At the Helm: Leading your Laboratory*, 2nd edition. Cold Spring Harbor Laboratory Press, USA.

7 How to Publish One's Results

Every researcher has to submit research articles and communications to journals for publication in the scientific literature. At the 27th European Crystallographic Meeting held in Bergen, Norway, in August 2012, there was a symposium aimed primarily at early career researchers. I presented some guidelines and advice in a lecture on 'How to Publish One's Results'. The chair of this session explained my qualifications:

> An experienced research scientist with extensive involvement in academic publishing, and had served as Editor-in-Chief of *Acta Crystallographica* and Chair of the International Union of Crystallography (IUCr) Journals Commission between 1996 and 2005; had also served as a Co-editor of the *Journal of Applied Crystallography* since 2005; had also been Joint Main Editor of *Crystallography Reviews* since 2007; and was the official IUCr Delegate to the International Council for Scientific and Technical Information (ICSTI, 2005 to 2014) and to the International Council for Science (ICSU) Committee on Data (CODATA), from 2014.

I covered the following topics:

- When is it timely to publish?
- Is there intellectual property (IP) that must be protected?
- Which journal to select?
- Which editor to select?
- To nominate referees or indicate referees not to use?
- The draft: Escalate the key points to the abstract or even the title
- Inclusion of the data, processed and derived data, and also, if possible in spite of likely large file size, the raw data
- The reviewing cycle
  - The need to listen to the referees and the editor
  - The revised version: Make it easy for the editor to accept
  - Proofreading: Read your proofs carefully; technical editing can introduce errors
- What helps to attract citations?
  - Reviews and books can help
- Wider issues and impacts
- Future publishing landscape?

One of the potentially most difficult steps in the article drafting is the authorship list itself:

- Which coauthors are appropriate to include or acknowledge?
- In what order should the coauthors be placed?

Journals have tried to be helpful to authors on this by introducing the *role of coauthor* definition such as who conceived the project, who did the research in its various portions (e.g. chemical synthesis, sample purification, physical measurements and/or analysis), who coordinated it and who wrote the paper? Suffice to say that only those directly involved in the work should be authors. With the authorship membership settled, then comes the authorship order. Different conventions exist in different fields. Old fashioned would be alphabetical by surname. As e-mail came in the asterisk against one name as the corresponding author became emphasised with the addition of that person's e-mail address being given as well. This seemed to coincide with more collaborative research between laboratories, bringing different research specialisms together, and so the introduction by journals of two corresponding authors, which is appropriate I think due to the distinct research specialisms of each laboratory. This would also quite possibly bring two, rather than one, prominent persons in the execution of the research, and so the first and second author names' position would each also be with a superscript saying something like 'contributed equally to this work'. Overall, these issues can be ultrasensitive, not least as people's careers can depend on settling this properly and fairly. Making progress will critically depend on the good handling by the lab leader(s) involved in coordinating the research.

After an editor receives the manuscript, he or she may well reject your paper without sending it out for peer review. Authors often fail to appreciate this and can get annoyed because their paper has not been reviewed. However, many of the high-impact journals may only accept as few as 5% of the papers that are sent to them; although more typically, say, 25% are accepted. To achieve a high impact, they are incredibly selective. To get to the refereeing stage, your cover letter explaining the high significance of your paper, and hopefully also its interest to more than one specific research community, will need to be very carefully thought about by you and your coauthors. Also, if your cover letter is in fact unconvincing to you yourself, then you better not waste time on the specific article formatting required for that particular journal. For your research to reach a wide audience via a high-impact journal will be highly regarded and is a good thing for your career development; although with the growth of Google Scholar, the journal brand, is becoming less important in the discovery of research. Also, in the specific research areas relevant to medical research, *PubMed* is a service of the United States National Library of Medicine that provides free access to MEDLINE, the National Library of Medicine database of indexed citations and abstracts to medical, nursing, dental, veterinary, healthcare and preclinical sciences journal articles.

So if a paper gets through the editor's screening stage, it will go to reviewers who are anonymous. Sometimes, with, for example, medical research results, reviewers will also not know who the article's authors are either; this is called *double blind refereeing*. Different journals have different methods of reviewing. Most commonly, journals will send each paper to two scientists who are knowledgeable in subjects relevant to your article, who will review it and then provide written feedback to the editor with a recommendation. Of course, two reviewers may have different views on the same paper, so sometimes the editor will need to send the article to a third reviewer. Once the editor has received the views of the reviewers, he or she will respond to the author with a decision. The decision may be to accept the paper straight away, but that does not happen very often. More commonly, the recommendation provided by the reviewers is that the paper requires minor amendments, which the author then makes and resubmits. If the reviewers' comments are taken into consideration and necessary changes are made, with a point-by-point response by the author to the reports, these papers are then most often promptly accepted for publication. When the reviewers suggest that a paper needs major amendments for it to be publishable, it might well be difficult for you to know what to do next. Obviously, you need to carefully consider these major amendments, and you may perhaps consider summarising your revision plan in written discussion by e-mail with the editor. This at least makes the best state of preparation for your revisions before the editor goes back to the reviewers. Assuming your paper is accepted, the editor will send it to the publisher who will send you a proof of your article as a final check, after which your article is eventually published. Alternatively, it is possible that your paper is rejected. My advice in such a case is that you learn from the feedback, undertake more experiments or calculations as may be required, and then when ready try again, most likely elsewhere, as some journals have rules against resubmission, although others have no objection. Overall, you should accept that a significant number of your papers are likely to be rejected during your career.

When is it timely to publish? Firstly, has your research actually reached any conclusions? Secondly, is there IP that must be protected? Basically, you should not reveal your results in public by any medium (talk or publication) if you think you might wish to seek a patent. Most universities have IP officers to help and advise you. Thirdly, the more judgemental is whether to publish a preliminary or a fast communication to establish precedence; the Royal Society of Chemistry's *Chemical Communications* is one such journal that allows this option.

Which editor to select? Never pick a friend or a colleague; it is an easy way to risk losing your friend. Pick someone who gave you an impression of being fair and level-headed when you met them at a conference or heard them give a seminar. Also, by talking to your colleagues, they will likely recommend one to you who has a long-standing good reputation of being a fair editor in your field.

The article drafting: The whole text is important but getting the title and the abstract right is critical. You must be aware as you write of the possibility to escalate the key points and even revise the title! Note that there are abstracting services who will offer readers your title and abstract, and so this can often determine whether your article will be read in detail (apart from your friends and/or competitors). For the title, it is worth noting that researchers search online using the keywords they are interested in, and so including those in a title if you can is usually good for helping your research to be discoverable.

Increasingly, and a very good thing, a paper will be linked to the data that underpin what you are reporting [1–3]. In your research community, there will hopefully be an agreed view, for example, set by the professional organisation representing your community via its journals or by the relevant database(s) of the data reporting standards required to sustain a reported result. In my field, for example, there is a consensus of data standards in small-molecule crystallographic community results reporting. Thus, there are web-enabled tools to check the data such as the check the crystallographic information file (checkCIF), and the Protein Data Bank (PDB)* has launched a data validation tool to help the protein crystallography research scientist author. The introduction of automated quality assessment during the peer review of small-molecule and inorganic structure reports provoked a certain amount of controversy, but with careful tuning of the checks and the tests, and a sympathetic application of editorial policy, checkCIF has become accepted as a de facto quality indicator for this community. In the field of biological macromolecular structure determination, such a consensus is still being perfected. In any case, the access of the reviewer to the data underpinning an article is critical.

On editors: A general point to note is that they are not always as tough as their reputation might seem! Editors can help authors. The editor is of course the gatekeeper of good science, but you should expect them to be constructive, offering help and expertise that can lead to article acceptance after a major revision. As an editor myself, I could readily see that the text can often be improved, and I have seen that papers can be submitted even without a Conclusions section! A paper without conclusions is obviously not yet mature for publication.

Referees (reviewers) are vital to the process of judging a paper. As an author, you should listen to the referees and the editor, in the same way that when you are a referee, you would hope that your report is carefully considered and in good measure acted upon.

You will be interested to know what, in my view, are the usual reasons for rejection of an article:

1. It was sent to the wrong journal; that is, it does not fit a journal's aims and scope. Here you need to be careful to read the notes for authors and not the journal's advertising, which may well indicate a wider scope to win readers and subscribers than a journal would indicate to its potential authors via its Notes for Authors.
2. It was technically flawed in the data or the procedures used.
3. It fails to say anything of significance (makes no new contribution to the subject) or states the obvious at tedious length.
4. The writing has bad grammar and punctuation.
5. It is so long worded that the text exhausts the patience of the editor and the referees and would therefore, in the mind of the editor, exhaust the readers

---

* The PDB launched in 1971 is now a full-service database for the three-dimensional atomic structure data of large biological molecules, such as proteins and nucleic acids. The data, typically obtained by X-ray crystallography, neutron crystallography, nuclear magnetic resonance spectroscopy or cryoelectron microscopy and submitted by scientists from around the world, are open access via the Internet.

and who would never cite it! Conversely, of course, a too terse an article can be cryptic rather than explanatory.

6. Other reasons for rejecting an article may include the failing of ethics tests such as plagiarism checks (involving copying from your own or from other peoples' papers) or rubbishing or ignoring other peoples' work.

This list does not include impact assessment as a reason for rejection, as I have discussed that at length earlier in your choice of journal.

What helps to attract publication citations?

1. Try and secure the front cover of the journal issue in which your article is published. You would do this by sending a letter explaining why your article is significant and new and, therefore, be good for the journal. Naturally, you need to offer as attractive picture as possible for the front cover image.
2. Alert colleagues in the field by sending out a weblink to your article.
3. Use social media (Twitter and Facebook; LinkedIn also).
4. Link to your papers in your e-mail signature.
5. Discuss with the publisher's marketing team; publishers are increasingly working with editors and authors to find good papers to promote.
6. Use services such as Kudos that are available (see, for example, https://www.growkudos.com/about/researchers).
7. Talk about your work at conferences.
8. Prepublishing your paper at bioRxiv (http://biorxiv.org/) or arXiv (http://arxiv.org/) is an important way to put a marker in the ground in some research fields (having checked beforehand that your preferred journal does not object to this preprint publication).
9. Most important of all, publish good work!

Reviews and books are a whole special category of publication that you could write, and in which your work will no doubt feature but obviously needs to be an extensive coverage of the relevant literature. These publications allow you to bring together a theme of your work and naturally set your publications in the context of the field. To convince a publisher that you are qualified to write a book, you will need to have a good reputation in that field, and so this is very likely to be something you would consider taking on mid-career or late in your career. Alternatively, working with a senior researcher/academic in a field to coauthor a work is a good option if you are in an early career post. A review article of a research topic can seed a book research monograph, but copyright permissions would need your careful handling and negotiation unless the review was published by the same publisher as you now propose for your book. I have published several research books, as sole author, as a coauthor or as an editor, and I served on two book series approval committees, but I have not written a teaching science textbook and so refrain from offering advice to you on that category of science books.

Wider impacts, for example reaching the media via a press release, will increase your publication downloads, although not necessarily from your scientific colleagues. Alternative metrics, *altmetrics*, such as downloads, mentions on Twitter and so on are growing in importance and indeed often featured now on publishers' websites alongside articles. I discuss press releases in Chapter 25.

What will the future publishing landscape become? In my view, there will certainly be yet more interactive content – including embedded videos and/or audio within articles, as with one's electronic newspaper. Semantic enrichment will likely develop, for example, where a chemical is mentioned, it can be directly linked to the catalogue of a chemicals' supplier. Routinely, there will be more and more cases of linking to raw data sets underpinning the words of the article. Indeed, it is remarkable in my view that there are any science disciplines where data linking to articles was not deemed mandatory! There will be an even greater role for open access; a publishing revolution is indeed in progress, and for funded research, it is absolutely welcome, but for unfunded research, often the most adventurous as funding agency committees are quite often risk averse, there should not be financial barriers to authors to publish. This requires journal subscribers to come to the rescue to pay publishing costs for the category of unfunded research.

In summary, the overall quality of scientific argument in a publication depends on the quality of the underlying data, on the language and syntax being clear, on the rigorous analysis and on the direct accessibility of the relevant data. In my experience, it is very rare that my chosen referees would give the verdict on an article submitted to me as editor: *publish as is*! So peer review is going to be important, but unfortunately, not all journals provide you, the author, with the highest quality of peer review, and so you need to choose your journal carefully, and for good measure, send your submitted article to a couple of friendly experts in parallel with the journal's peer review. Overall, be guided by the principle that the primary purpose of publication is to share scientific knowledge, and the associated data, and thus, keep science going as a healthy and collective enterprise.

REFERENCES

1. J. R. Helliwell, P. R. Strickland and B. McMahon (2006) The role of quality in providing seamless access to information and data in e-science: The experience gained in crystallography. *Information Services and Use* 26(2006): 45–55.
2. P. R. Strickland, B. McMahon and J. R. Helliwell (2008) Integrating research articles and supporting data in crystallography. *Learned Publishing* 21: 63–72.
3. J. R. Helliwell and B. McMahon (2010) The record of experimental science: Archiving data with literature. *Information Services and Use* 30(2010): 31–37.

8 How to Communicate Your Results

While Chapter 7 dealt with publications, here I cover all other avenues of communication. In any case, a publication is not a two-way communication process; it is one way. Yes, there are citations of publications, but again it is not communication in the sense of conferring together with other people. A conference is a chance to confer. A speaker at a conference who leaves no time for discussion with their audience within their allotted time slot is working totally against the spirit, indeed point, of a conference, literally, the chance to confer with others. When I have chaired sessions at a conference I have been seen to even move someone off the stage if he or she starts to run into the next speaker's timeslot. This will have been preceded by my signalling, increasingly frantically if the speaker continues, that a speaker is using up the time that their audience will expect to be able to ask questions of, or make comments to, a speaker. There are quite a lot of different time allocations as a speaker that you might have to work with, but suffice to say, if you have accepted that task, then you are duty bound to stick within it and, indeed, leave a portion of time for questions and answers. A simple rule would be 10% of the total; so a 25-minute talk should include 2.5 minutes for questions and answers. In organising your talk, you need to do a practice beforehand, indeed more than one practice if you overrun on the first practice. Generally, you will have slides in PowerPoint or an equivalent program, and so to start with one slide per minute is a good guide to the total number of your slides. This also requires, secondly, each slide to not be overly detailed. Whatever you have included on a slide should be explained to the audience. If you have something you do not explain, then, to say the obvious, why have you included it?

The legibility of your slides will be vital! Obviously, you need to choose a clear font that is easily legible, even in poor lighting conditions. Certain colour combinations should also be avoided such as red typeface on a blue background, or yellow typeface on a pale green background. White typeface on dark blue is good, as is yellow typeface on dark blue. Bear in mind that a significant percentage of males are red/green colour blind, a colour combination therefore not to be used.

Overall, then, you are trying to get your results presented in a digestible form for the audience and without haste. Another way of looking at how to approach this is to ask what the barriers are to communication. Wikipedia offers a list of barriers of which I pick relevant ones:

- Physical barriers
- Ambiguity of words/phrases
- Individual linguistic ability
- Physiological barriers

This list is explained with examples. I will offer my own examples.

Physical barriers to communicating one's results can include a lecture room's poor lighting and/or poor acoustics, outside noise, laptop/projector problems, or no laser pointer.

Ambiguity of words/phrases can include use of colloquial expressions from your country being meaningless to an international audience.

Individual linguistic ability means you need to be mindful of where your conference is located, so, for example, if the audience is largely one with a different language from your own (conference languages usually being in English, this means non-native English speakers), you must not speak too fast, and the usual rule of not directly speaking the words on your slide I think can now be exactly how you would proceed.

Physiological barriers means clear diction and projecting your voice are very important. Quietly talking down to your chest can defeat even the best microphone! As you travel around the world, you may well find yourself presenting at a body time of 4:00 a.m., when locally it is noon. Adrenalin will probably get you through this, but in the question-and-answers stage, you can always say 'Please repeat the question' to avoid saying anything unduly hasty in reply. Travelling from the United Kingdom to the Far East such as Japan, I found jet lag a special challenge. The best option I found, to adapt as quickly as possible to local time, was to arrange to give a lab seminar late in the afternoon of the day of my arrival; flights would arrive 9:00 a.m. in local time. That way, the adrenalin would see me through my talk and prevent me from going to sleep in the afternoon, a bad mistake if I wanted to get a night's sleep during my visit! Also, the locals would take me out afterwards for a nice dinner, which would see me off to my hotel bedroom in a good glow ready for sleep.

Another barrier to your audience following your answers to their questions is if you were to reply in a verbose manner and/or try and make some other point than what was asked about. You need to be clear and succinct; sadly, you will indeed witness a speaker who breaks this simple guideline. To say 'I don't know' is perfectly fine! To say 'That's a very good suggestion/observation' is fully in the spirit of conferring together. If the questioner tries to put you down, which sadly can also happen, you can always opt for 'I don't agree with you on that'; neat, simple, succinct. You could also add, if appropriate, 'That probably requires a lot of detailed discussion which I am happy to do over coffee'. My point being to keep the interaction with that person as civil as possible.

I wish to give you an example of a lecture I have given, including a video of it and my slides. It is rare that one's lectures are recorded, but there is one type of lecture that does attract that level of support and interest, a prize lecture. I was awarded the Patterson Award of the American Crystallographic Association in 2014 and the Max Perutz Prize of the European Crystallographic Association in 2015 for my research and development involving synchrotron X-ray, and neutron, protein crystallography instrumentation, methods and applications. Each of these lectures and slides are available via the weblinks in the American Crystallographic Association [1] and European Crystallographic Association websites [2]. As objective measures of my lecture and research content quality, comments from student bursary recipients who attended my Max Perutz lecture, for example, included 'An overview of his exceptional work'. 'Inspiring work on synchrotron crystallography'. And 'An amazing talk'. So I hope these two research-focused lectures serve to illustrate the practical points I have made to you in this chapter about communicating your results. I can add as specific details of finalising them that each lecture took around six months of preparation including multiple practices (about five times) and included my wife and work colleague, Dr Madeleine Helliwell, listening to each practice and so helped me gradually refine it. I also gave a practice presentation of my, nearly final, Patterson Award lecture to colleagues in the University of Manchester, Institute of Biotechnology. I am especially grateful for all the constructive advice I received. Especially challenging was to be sure of the time each lecture would take me to give it, and to ensure that I stayed within the time allocated to me. Naturally, each practice, even of the same slides content, would have a slight variation (2–3 minutes for a total target time of 45 minutes) in the time I would take. I also worked on understanding which slides or time of day caused the variation in total time. This degree of perfection in preparing lectures is obviously unusual.

REFERENCES

1. American Crystallographic Association (2014) Patterson Award of the American Crystallographic Association in 2014. Accessed 7 August 2016. Retrieved from http://www.amercrystalassn.org/h-helliwell_award.
2. European Crystallographic Association (2015) Max Perutz Prize of the European Crystallographic Association in 2015. Accessed 7 August 2016. Retrieved from http://ecanews.org/mwp/blog/ecm29-opening-ceremony-2/.

9 How to Manage Your Time

> To achieve great things, two things are needed: a plan and not quite enough time.

> **Leonard Bernstein**

Time is a very important commodity, especially your time and those of your staff! I learnt about the time management quadrant (Figure 9.1) when I was the director of a large department in the scientific civil service (i.e., as a government-employed scientist), at a standards of management excellence course. It is simple but so effective.

In universities, I had so often felt I was in the top left quadrant either fire fighting (dealing with the unexpected or the unwelcome) or chasing deadlines. Such fire fighting was, it seemed, hardly ever my fault, rather they were problems caused by others. One always tried to deal with the unexpected or the unwelcome of course, but the trick is to try and ensure that a similar situation does not repeat itself. The lesson I also learnt on this scientific civil service course was that one should try and spend most of one's time in the top right quadrant, the *important but not urgent* quadrant. This fitted not only the role of the director but also, I realised, the role of a laboratory leader in academia! A classic, self-imposed situation, for me as an academic, was leaving a grant proposal too late to complete it without a rush. Now, I know how to avoid that; start early enough to avoid a rush, and if needs be, set oneself an earlier deadline than the funding agency's formal deadline!

One of the gradually increasing requirements for a successful research grant proposal is sufficient preliminary data to convince the funding agency's committee members that the risks of failure in the work are well contained. Of course, there will, or should be, some risks about a proposal, otherwise the work is going to be so unadventurous as to not get funded either. So in planning your schedule for your grant proposal to be submitted by a given date, you must start planning the slowest-moving item. A bit like on a long distance walk or a bicycle ride, you have to plan by the weakest person in the group taking part.

Quite often though, when making a list of to-do items, I would notice that the not so important, whether urgent or not, were the most attractive and pleasurable! This sort of thing requires a real sense of self-discipline and strength of character that you simply have to steel yourself for and focus on the important items. A simple method, having made your list, is to put them in ranking order.

As a mentor to various people over the years, especially those at a career ceiling, one needs to invite the person, the mentee, to develop new insights into their mode of working. A time analysis can be informative! Write down the areas of your activity and over a significant period, say a year, estimate the percentage you spent on each activity. Check this overview for the last year with typical weeks, for example, in a

Important/urgent	Important/not urgent
Not important/urgent	Not important/not urgent

FIGURE 9.1 The time management quadrant.

teaching semester and out of it, just as a reality check. Ask yourself, 'Is this the best use of my time in the various tasks?' Are any activities missing that you should make time for? One thing I noticed quite starkly between the scientific civil service and an academic scientist was that in the former, general skills training was compulsory, and in academia, it is voluntary. There are certain exceptions now; if you are to be on an interviewing panel, the university requires you to have undertaken equality and diversity training first. Basically, there are legal dangers for the university if an interview goes wrong. But the university does not seem to care too much about time management training; you seem to be expected to find time, whatever damage it does to your home life. The fact is, without any conflict, you can be more efficient. There are still limits, of course. Individual university departments now may have a workload model to try and ensure the fair distribution of tasks between different academic staff.

As you become more well known, you will receive invitations to undertake many types of service to the community task. Choosing what to say *yes* or *no* to is a skill worth developing. There will also come various, indeed many conference invitations. Overall, how you develop your approach on this will depend on your energies and appetite, as well as taste. As will become clear through this book, I have tended towards saying *yes* rather than *no*, and that perhaps reflects my broad outlook and a collegiality of approach! In the case of conferences, what you can accept will depend on the conference budget or more likely your own budget. In all cases, you must be self-critical; in order to do a good job if you accept you need to consider the following:

1. Are you competent to undertake the task that you are being asked to do? Should you explain caveats on what you can do and what you cannot do in the task envisaged for you?
2. Are you already undertaking a similar task to the new one you are being invited to take on? If so, you are already 'doing your share'/'gaining that type of experience'. So you can just say that you are already fully committed at this time.

3. For conferences, it is important to not only consider your budget, as I mentioned earlier, but also ask yourself if your diary and/or workload at that time already is congested. It is also important here to remember your work–life balance.
4. Invitations that are a long time ahead are particularly difficult to decide upon as your diary may well be clear while you are thinking over an invitation, but you should also look ahead as to when the major conferences in your field are planned, and they can indeed be planned with dates known three, even four, years ahead.

10 How to Use a SWOT Analysis to Good Effect

Strengths, weaknesses, opportunities and threats (SWOT) analysis I have found is a useful tool for gaining new perspectives in mentor–mentee discussions when I was helping a person who felt they were at a career ceiling. To assist such discussions, and to show my own belief in such an analysis, I would prepare my own career SWOT analysis and share it with my mentee. This also created a convivial atmosphere for discussion between us on what was clearly a most serious topic, career progression frustration for the mentee. Obviously, such discussions in the details were completely confidential, and I cannot share with you any case study examples although general principles did emerge. Firstly, it was most likely that the analyses could be centred on the person's own details being the dominant factor. Secondly, however, the analyses could reveal a dysfunctional work environment in the mentee's department. This latter problem, in assisting the mentee work towards a solution, would involve my sharing of my own department's documents for, for example, annual appraisal, and which I deemed to be constructive in their review of the previous year and the setting of objectives and skills training for the coming year. The primary aspect, and most common though, was that the person's own details were the dominant factor and each time led to an overhaul of their career strategy.

So, what is SWOT? I offer you my own personal example that I prepared for a mentor–mentee discussion (prior to my formal retirement in 2012) listed below.

My own career SWOT analysis in May 2011:

Strengths

A broad coverage of science formal qualifications and skills such as professor of structural chemistry (since 1989) and a DSc in physics (1996; York University) and also FInstP, FRSC and F Soc Biol.

Nice highlights on my curriculum vitae e.g. Banerjee Lecture Silver Medal (first) presented in Calcutta, India; British Crystallographic Association Lonsdale lecturer in 2011.

Senior management experience e.g. Council for the Central Laboratory of the Research Councils (CCLRC) director of Synchrotron Radiation Science (and head of the Daresbury Laboratory Synchrotron Radiation Source, which comprised 240 staff and an annual budget of ~£20 million); editor-in-chief of *Acta Crystallographica* from 1996 to 2005; and president of the European Crystallographic Association (2006–2009).

An h-index of 35 for my publications.

Weaknesses

Extremely busy with several time-consuming but important administration tasks (Department of Chemistry Gender Equality champion; senior mentor for new academics), as well as needing to prepare more and more research grant proposals as success rates have dwindled to 1 in 6 typically, sometimes to only 1 in 8.

Opportunities

Buoyant range of research proposal ideas.

Research active both personally and with a buoyant research group (i.e. with one or two final-year undergraduate project students and funded PhD or self-sponsored MSc students).

Threats

The UK Government budget for universities and the UK Research Councils and the EU budgets in the coming years may decrease in real terms.

Since then, I have retired; well in fact, I call it *semiretired*, although I am now on a pension, not a salary. So, for this book, I have prepared an updated SWOT analysis (listed below) for my semiretired state with respect to research.

My own career SWOT analysis in December 2015:

Strengths

A broad coverage of science formal qualifications and skills such as professor of structural chemistry (since 1989) and a DSc in Physics (1996; York University) and also FInstP, FRSC and F Soc Biol;

Highlights on my curriculum vitae e.g. Banerjee Lecture Silver Medal (first) presented in Calcutta, India; British Crystallographic Association Lonsdale lecturer in 2011; Patterson Award from the American Crystallographic Association in 2014; Max Perutz Award of the European Crystallographic Association in 2015; admitted as a fellow of the American Crystallographic Association in 2015.

Senior management experience e.g. CCLRC director of Synchrotron Radiation Science (and head of the Daresbury Laboratory Synchrotron Radiation Source, which comprised 240 staff and an annual budget of ~£20 million); editor-in-chief of *Acta Crystallographica* from 1996 to 2005; and president of the European Crystallographic Association (2006–2009); main editor of *Crystallography Reviews* (2012–).

An h-index of 43 for my publications.

I have much more time now to update my research skills (basically, my computational rather than laboratory bench skills).

Weaknesses

To remain research active personally, i.e. without funding or research students.

Although a lab leader these last 23 years, from 1989 to 2012, I am now no longer a lab leader. So I need to guard against causing any difficulty for the leader of new projects I am working on as a participant. By this, I mean by me taking any actions as if I am still a lab leader!

Opportunities

Continue my scientific research discoveries.

Support my past PhD students and PostDocs in their career progression.

Continue my science editorial work.

Continue my formal representation roles for the International Union of Crystallography.

Engage with social media such as Twitter to tweet about science discoveries I admire and my own analyses about science policy changes.

Threats

The open access to the public of publications and data from funded research is obviously a good development, but as an unfunded research author, I increasingly rely on subscribers to journals to cover my research publication costs. At present, I can still publish my research without article processing charges in my learned society, not-for-profit, journal titles, but this may change.

Any sudden decline of my good health as I get older is an overall threat.

A set of helpful tips to create your own SWOT career analysis, which I have adopted from the Slideshare archive [1], is shown as follows:

Strengths

What skills or experience do you already possess?

Give examples of how you have used your skills.

What do you do well already?

What do other people perceive as your strengths?

Weaknesses

What areas of your development could you improve on?

Do you lack experience that you may need for your long-term career?

What do you sometimes do poorly?

Opportunities

What activities and opportunities are available to you in your learning and development?

By considering your strengths, do these open up new opportunities for you?

By considering your weaknesses, could you identify opportunities that would eliminate them?

Threats

Are there external environment factors that may badly impinge on you in the future?

Are there things you have failed at in the past that create new threats or barriers to you now? (These may well be handled within the third point I list under opportunities, or you may have to work around this threat that you yourself have created.)

Overall, if you imagine that you need to undertake a SWOT analysis, I would commend that you register for your institution's mentor–mentee scheme as soon as possible, or if that does not exist, contact your professional scientific society, which may also offer such a mentoring scheme. Then, with your newly found mentor, you can work together to undertake SWOT analysis as carefully as possible for your career health check.

REFERENCE

1. Nottingham Trent University (n.d.) Helpful prompts to complete your SWOT analysis. Accessed 7 August 2016. Retrieved from http://www.slideshare.net/CareersRish/personal-swot-analysis-example-28527317.

11 How to Use Social Media for Your Work

In November 2012, I was at the International Council for Science committee on data (CODATA) biennial conference, in Taipei, as the incoming representative for the International Union of Crystallography. I entered the breakfast room, and being on my own, asked if I could join someone, who was on her own as well. She was checking her morning list of tweets on her smartphone. I resisted making some critical comment, as I did have a bias against Twitter, which I seemed to have heard about in some fairly derogatory ways, although I could not quite recall where. She was the delegate representing CODATA Canada. I enquired about Twitter. She explained how useful it was for her work. 'How do I start?', I said. Well, you register your name and follow some people, and they will more than likely follow you back, especially if your short summary about yourself resonates with their stated interests. So that was the start, me being sociable at breakfast then registering my Twitter username. Being in Taiwan I suppose, when I entered John Helliwell, my Twitter name became *@HelliwellJohn*. I took considerable care over my Twitter account descriptor (Figure 11.1). So far, I have updated my photo about every year. As a theme, and to be distinctive, I have had photos of me in some cycling expedition, a sort of compromise between my descriptor, which is work, and my picture, which is me relaxing.

Within a year, I was following 1000 or more people and organisations. I had tried to be disciplined about it, but following a large number of diverse work people and work organisations plus some sports people, I had so many tweets per day I could not possibly read them all. Over a week, I deleted about 100 per day, solely focusing on my work area and got down to just under 200 that I was following. It has steadily grown these last three years to around 700. Similarly, I am followed by about 700.

Twitter, I have found, is an incredibly powerful tool to learn about what is going on in science in general and my research field of crystallography in particular. It is rare that I do not check my tweets each day. It is my personalised work daily news.

I have listened to presentations from users of Twitter by, for example, other journal editors. For example, I heard from two such in talks they gave one after the other. One said, 'tweet regularly and often', and the second said exactly the same thing, 'tweet regularly and often'! But each had given their number of tweets; one as 15 tweets per day, and the other was one per day. I made the comment that 15 per day meant that person, that journal Twitter account, was what I had started calling a Twitter pest creating considerable clutter in the stream of tweets that I received.

Besides those who tweet regularly and often tweeters, who I unfollow even if they occasionally have a good tweet, there are also the hobby horse tweeters. Although their tweets are work in their content these tweeters are one-theme tweeters, more or less. Well, one or two tweets from them is enough, I often find, although not always.

FIGURE 11.1   My *@HelliwellJohn* descriptor. Research Crystallographer based at Manchester University before at Daresbury Laboratory. Journal Editor. Author of science articles and several books. Educator.

There are the tweeters who in among a work tweet will tweet about what they have for breakfast, what their cat is doing etc. I can tolerate a certain amount of that if they have other more interesting tweets, but unfollowing is a likely possibility; i.e. more unnecessary clutter, which must be removed.

One of the worst of all sins in tweeting I think are the swear word tweeters. Is that really necessary? Then finally, worst of all, are the 'abuse others' tweeters, who I find really strange and are absolutely immediately unfollowed.

So there are hazards of Twitter, but the simple guidelines mentioned earlier will allow you to decide how to fashion your own tweeting to your personal taste. It is the new era of developing your social media skills!

Are there nonwork tweets I would like to send? I do get tempted to tweet about politics especially when, only occasionally, tweets I receive stray into science budget cuts as their topic. While my tweets might assist in an analytical way, I might hope, in general, the rule for parties of 'don't talk about politics or religion' is a good one I think.

So, carefully managed, in content and the time that you spend checking your tweets, I am definitely a convert to the use of Twitter for my work.

What about other social media? Facebook, Academia.edu etc. I am registered, but I rarely use them. LinkedIn I have found is occasionally useful, for example, when I am trying to find someone's e-mail address. If I were searching for a job, it would then no doubt come into its own I imagine.

Google Scholar is incredibly useful. I have made my publications list via my Google Scholar profile public. I receive e-mails from Google Scholar telling me when one of my publications has been cited, although it could be improved if their e-mail would actually say which paper had been cited. I can also see at a glance my current h-index or the total number of citations detail, which are occasionally needed. It is important to check what publications Google Scholar holds in their list for you; there is a John F. Helliwell, an economist, and whose papers occasionally surface in my own list, and I have to weed those out and any other stray entries, which I do when I check my Google profile, usually once or twice per year.

Research Gate is also a useful check of citations to my publications. It presents the hazard, however, of one's full texts getting into Research Gate's archives e.g. from my coauthors who overenthusiastically upload them, not thinking of journal copyright that may be contravened. Research Gate now also accepts data sets, a very good development, at least as a temporary measure until journals start attaching data sets or universities start providing data repositories and obtaining the digital object identifiers for each data set. The University of Manchester is one of the very first that I know of to provide a data repository for its research staff, so I am privileged to be so well served by my data librarian. Research Gate also provides a forum for research questions and answers. As I am well served in my research area by a tailored set of bulletin boards for the software I regularly use, I have not made much use of the Research Gate question and answer role; there is real scope here though for improving my engagement with what Research Gate offers.

This discussion of my preferences for use of social media are illustrative. For your own engagement with social media, I would encourage you to explore your options, albeit with the occasional cautions I mentioned earlier. Social media options, and other similar web tools, are a fast-developing area but with incredible potential for your scientific life, as an educator and a researcher. You will also, with Twitter, learn the skill of making your point or passing on information in a tweet of 140 characters or less!

12 How to Avoid the Travails of a Research Manager (or the Pitfall of Ending Up Not Doing Science)

By prevailing over all obstacles and distractions, one may unfailingly arrive at his chosen goal or destination.

Christopher Columbus
Italian explorer to New Worlds

It was not until my early fifties that I realised that I had become a research manager. It had happened gradually. I was running a lab, teaching undergraduates, supervising postgraduates and undertaking at least one senior administrative task in my department (chief examiner, if I recall the time precisely, which I did for three or four years). I realised that my science skills were out of date, be they at the lab bench or in calculation. To define a precise moment is difficult, but suffice to say that I was uncomfortable about it. There were many job satisfactions of bringing the best out of others moving forward with new discoveries and seeing students get their degrees, be they BSc, Masters or PhDs. But something was missing. By coincidence, there was a less and less chance to win research grants; one in three had become one in four had become one in five. My research grant proposals were, in my view, every one of them was absolutely worthy of funding but the odds began to stack up for me as for everyone else. Sure, we all had 'fisherman's stories' of the grant that should have absolutely been funded, but now, there was a lottery element to it all. To survive, I would have to get back to doing research myself; it simply was not the size of group to keep myself fully occupied managing it. This cloud had a silver lining; I would rediscover the joys of my own work. But rekindling my detailed skills, the skills I used to be fluidly proficient at and now also the new ones would be a challenge. But a regeneration process began for me.

What were the practical steps of the regeneration?

First, I took a research sabbatical; the department had a proactive sabbatical policy, and I had the necessary number of full years of service since my last sabbatical. My number of years service calculation was a little bit more complicated than usual as I had lengthy periods of service (full time or part time) at the nearby Daresbury Laboratory working for the Science and Engineering Research Council (renamed later as the CCLRC, and later yet another name, the Science and Technology Facilities Council). Suffice to say, I certainly had enough equivalent years of service to justify my application for a research sabbatical, accompanied of course by a research plan including the development of my research skills. Being in a highly privileged research and education environment in the University of Manchester, I decided, unlike my sabbatical in 1994 first to Cornell University and then to Chicago University, to take this locally. I embarked on relearning electron microscopy (EM) in the next-door biological sciences department with my host, Dr Clair Baldock, working on a key aspect of the coloration of marine shells (specifically, lobster carapace), and within the carapace, the one-third of a million molecular weight multiprotein and carotenoid complex alpha crustacyanin. I say relearning in that as my undergraduate physics final year project at York University, I had done an EM project on crystalline thin metal films. Although very, very different samples to a lobster shell protein nevertheless I did at least know one end of an electron microscope from another! In my research area of X-rays and neutrons as probes of the structure of matter, I had closely related experience, and so it was timely to reactivate my expertise and interest in electrons. Not least, the instrumentation had moved on a lot. Also for the software, although new to me, I would be under the guidance of an expert group of people! Clair and I by good fortune were approached by a physics graduate from Imperial College London, Natasha Rhys, keen to get involved with biophysics in general and molecular structure determination in particular, and with her own funds. We prepared new samples, measured many single particle images in the EM, and with a high-resolution crystal structure of a part of the complex from an earlier research I had undertaken with a PhD student, now Dr Michele Cianci, we obtained a negative stain image of the full alpha crustacyanin. This image albeit at the low resolution of the negative stain method provided the preliminary results for a high-resolution cryo-EM grant proposal, a necessary prerequisite of such proposals. Alas, the proposal was graded 'internationally excellent research but insufficient funds available', all part of the diminishing odds of success for grant proposals. But I was considerably the richer in terms of my skills, we had a nice publication on the negative stain structure, and very importantly, Natasha graduated with her MSc and went to Leeds University to undertake a PhD in the nanobiophysics research group of Dr Lorna Dougan there. (Natasha is now doing postdoctoral research at Oxford University.)

In a second domain of retraining in my sabbatical, with my freedom of time and free of regular work commitments, I set about refreshing my core expertise in calculations. In my field, a key annual school/training/conference event is the CCP4* protein crystallography study weekend. I started reattending these; the first one happened to be very timely being in the first few months of my sabbatical. I felt slightly guilty taking up a place that might have been occupied by a PhD student, but no one commented. But anyway, quite a core of experienced people besides the invited speakers and the large number of students do attend these study weekends.

Thirdly, I also set about checking out conferences that were coming up through the year, which had so far been on my *it would be nice to attend conference events* list. Naturally, this would mean attending conferences where one was not giving a

---

* CCP4 is the UK Research Council's Collaborative Computational Project No. 4 'Software for Macromolecular X-Ray Crystallography'.

talk, and I would have to pay for the registration, travel and subsistence costs out of my general lab funds, there being no funds attached with my sabbatical itself. My general lab funds were running pretty low, and so I committed some of my own money to this. My top priority for the use of these scarce funds was the 'European Charge Density Conference V' that I attended on the shore of Lake Como in Italy. Next on the list would have been the US Biophysical Society Annual Conference, but I just could not afford that.

Fourthly, since I was going to update all my calculations software, and since my then laptop was three years old, I decided on updating of my laptop hardware as well. The performance of laptops had, as we all know, progressed incredibly, and so even after three years, the computer performance could be transformed with a new one. This was so much so that personal computers in the lab were, for the sort of analyses I was undertaking at least, superseded by my new laptop's performance.

Overall, I came back to my regular academic year's duties thoroughly refreshed. The sabbatical is a wonderful institution of academic life and from which a concerted development of one's lab and analysis skills can be transformed. This was in addition to the domain of my research expertise now including two new research areas, EM of biological samples and ultrahigh-resolution electron charge density using X-ray crystallography. Furthermore, as the title of this chapter states, I had got myself out of the difficulties of, and moved beyond, being solely a research manager. Now, I could apply my own research skills hands on once more, which also brought back a personal sense of enjoyment to my research.

Of course, quoting Christopher Columbus, I should make clear that he set off to find a new route to India and ended up in the Americas, and there apparently he was not the first, being beaten to it by the Norseman Leif Ericson some 400 hundred years earlier! This just goes to show how difficult research and discovery can be, and also, suffice to say, his manager, if he had one, would not have known the route to India either.

13 How to Coexist with Competitors

The world has treated me very well, but then I haven't treated it so badly either.

Noel Coward (1899–1973)
Sheridan Morley: The Quotable Noel Coward (1999)

I imagine that like most scientists, at one stage or another of one's career, one finds oneself in a competitive research situation. In one such, I contacted the other research group leader, and we readily moved to a back-to-back publication of both sets of results [1,2]. There was no angst between us.

In the next chapter, Chapter 14, I will outline how one deals with one's mistakes including the situation where another team of scientists finds fault with a publication that you have. Such corrections by others might, most likely, be for the good of science, but it might be due to a competition of some sort, for funding, for example.

I am reasonably well read on at least some of the classics of competitive science:

• The race for the three-dimensional structure of DNA
• The race for completing the human genome by the public consortium and by a private company
• The race between the United States and the Union of Soviet Socialist Republics in the space race, or rather the race into space

There are also the competitions for priority over discovery:

• Newton versus Leibnitz for calculus
• Darwin versus Wallace for the theory of evolution

A less known controversy is that of whether Einstein should have cited Poincaré and or Lorentz in his 1905 special theory of relativity paper [3], which is a fascinating aspect of the biography of Einstein by Abraham Pais [4].

Referring to a recent report by the Nuffield Foundation, I can assert that times have not really changed, and from which report I quote,

Competition: High levels of competition for jobs and funding in scientific research are believed both to bring out the best in people and (yet) to create incentives for poor quality research practices, less collaboration, and headline chasing [5].

So this difficult state of affairs, now spanning all the centuries since modern science began, is carefully analysed by Fang and Casadevall [6] building on the classic work of Merton published in 1957 [7].

Contrary to the conclusion in the report by the Nuffield Foundation [5], Einstein was privately employed in the Swiss Patent Office; research funding was not at stake, and it was a pencil-and-paper level of resource activity anyway, as was the Newton versus Leibnitz affair. So one must conclude that human nature is at the core of the behaviours of these scientists. Therefore, I come back to personal skills; in this case, it is one's strength of character, ideals, vision and personality to behave well and keep a clear mind in trying circumstances that is at the core of how to approach such situations when one is confronted with them. My conclusion as to whether human nature can be trained into good character was not supported by Merton [7]. In my view, it is in essence a question of one's ethics and with one's peace of mind; I will cover these aspects in more detail in Chapter 31.

As a DPhil student in Oxford, I was really struck by the approach taken by Dorothy Hodgkin to her competitors in China on the research on the structure and function relationships in insulin, without whose normal functioning would result in diabetes. In the lab where I was based, I was told by a close colleague, Dr Guy Dodson, and I trusted Guy absolutely, that Dorothy Hodgkin stated the view in a visit there as to 'how much better she thought their research was'. Dorothy Hodgkin was awarded the Nobel Prize in Chemistry in 1964 'for her determinations by X-ray techniques of the structures of important biochemical substances'. She was also a tireless moral crusader via Pugwash as its president for over many years to remove the threat of a nuclear holocaust. I was very lucky and privileged to have interacted with her during my DPhil research studies at Oxford University. As I mentioned earlier, my

FIGURE 13.1 Here I am revisiting my DPhil graduate centre student residence Balliol and St Anne's Colleges, Holywell Manor, in 2014. I think I look quite content, albeit now with the direct knowledge of how science can turn competitive, i.e., one has to just take these things in your stride. I was a DPhil student, living there from 1974 to 1977.

supervisor was Dr Margaret Adams of Somerville College, and her DPhil supervisor was Dorothy Hodgkin. I revisited my DPhil student residence (Balliol and St Anne's Colleges Graduate Centre, Holywell Manor) in 2014 (Figure 13.1) when I attended the Resonant Elastic X-ray Scattering conference to present a keynote lecture. I was a DPhil student living at Holywell Manor from 1974 to 1977. As a DPhil student I had no direct knowledge of how science can turn competitive; one has to just take these things in your stride when they finally do happen, as they did.

With that said I think, while there have been competitions, and how to handle them I have not found easy, I have instinctively followed the quotation of Noel Coward at the start of this chapter. Actually, it was not only instinct. In those rare, trying circumstances, I would also recall my mother's saying 'as one door closes another one opens'; i.e., in any competition, you might lose out. So, learn from it, and wait for that other door to open, as it surely will.

REFERENCES

1. M. Cianci, P. J. Rizkallah, A. Olczak, J. Raftery, N. E. Chayen, P. F. Zagalsky and J. R. Helliwell (2001) Structure of apocrustacyanin A1 using softer X-rays. *Acta Crystallographica Section D-Biological Crystallography* D57: 1219–1229.
2. E. J. Gordon, G. A. Leonard, S. McSweeney and P. F. Zagalsky (2001) The C_1 subunit of α-crustacyanin: The de novo phasing of the crystal structure of a 40 kDa homodimeric protein using the anomalous scattering from S atoms combined with direct methods. *Acta Crystallographica Section D-Biological Crystallography* D57: 1230–1237.
3. A. Einstein (1905) Über einen die Erzeugung und Verwandlung des Lichtes betreffenden heuristischen Gesichtspunkt [On the electrodynamics of moving bodies]. *Annalen der Physik* 17(6): 132–148.
4. A. Pais (2005) *Subtle Is the Lord: The Science and the Life of Albert Einstein.* Oxford University Press, Oxford.
5. The Nuffield Foundation (2014, December) The findings of a series of engagement activities exploring the culture of scientific research in the UK. Retrieved from http://nuffieldbioethics.org/wp-content/uploads/Nuffield_research_culture_full_report_web.pdf.
6. F. C. Fang and A. Casadevall (2015) Competitive science: Is competition ruining science? *Infection and Immunity* 83(4): 1229–1233.
7. R. K. Merton (1957) Priorities in scientific discovery. *American Sociological Review* 22: 635–659.

14 How to Deal with Criticism

Jack Worthing: You are quite perfect Miss Fairfax.

Gwendolyn Fairfax: Oh I hope I am not that. It would leave no room for developments, and I intend to develop in many directions.

Oscar Wilde
The Importance of Being Earnest, 1894

Anyone who has never made a mistake has never tried anything new.

Albert Einstein

As Oscar Wilde so aptly put it, we can all recognise I think that none of us are perfect and can always 'develop in many directions'. Of course, it is one thing for Miss Fairfax to parry the warm compliments of a wishful suitor; it is quite another to be on the receiving end of criticism about one's research. In any case, as Einstein observed, you should not be afraid of trying something new for the fear of making a mistake.

There are of course two types of criticism that one may receive, constructive or destructive, or a mix of the two. As educators, with students, we are often in the position of dispensing criticism, which one surely aims to be constructive. However, one does read about student–supervisor relationships not working or breaking down. It is for which reason I imagine that one should not become friends with students. One should always remain at a distance.

A very common situation where you will receive and certainly must be prepared for criticism is reports from referees on your submitted research grants and articles. These aspects I have dealt with in Chapters 4 and 7. But, suffice to say, constructive advice and comments can be very helpful and lead to improvements in both types of submissions. In the case of destructive criticism, do not be afraid to politely say what you think is unhelpful.

A much rarer situation is where your published research receives a critique. The overriding aspect to not lose sight of if a critique comes in your direction, and if the critique of your research has really found a mistake in your research, then the response you should make, as matter of factly as possible, is simply to acknowledge the mistake. As a source of examples, the journal *Nature* has, firstly, on a weekly basis, corrigenda or errata from authors acknowledging and correcting errors in their previously published articles. Secondly, less often, but fairly regularly, published more in the front portion of an issue of the journal, are short statements from readers about an article and a response follows from the authors. Thirdly, the back pages occasionally, but rarely, have examples of an article's withdrawal. One of the most dignified, clear and matter of fact of these I have seen was by Wang et al. [1], which is surely an exemplar of good science and good scientific behaviour. It also took quite a time, over five years, between the publication of the article and its retraction.

Good science is going to involve the correction of mistakes. It is part of the normal scientific process. In an amazing dramatization of the process of science, Carl Djerassi and Roald Hoffmann wrote *Oxygen: A Play in 2 Acts* [2]. Quoting from the publisher's description of the play:

> What motivates a scientist? One key factor is the pressure from the competition to be the first to discover something new. The moral consequences of this are the subject of the play 'Oxygen', dealing with the discovery of this all-important element. The focus of the play is on chemical and political revolutions, as well as the Nobel Prize, which will be awarded for the 100th time in 2001. The action takes place in 1777 and 2001; and the play is written for 3 actors and 3 actresses who play a total of 11 characters. The world premiere took place in early 2001 in San Diego, and the German premiere was in September of that year.

Authors Carl Djerassi and Roald Hoffmann knew intimately what they were writing about as top flight scientists themselves; Carl Djerassi being known as the 'Father of the Pill' and Roald Hoffmann as recipient of the Nobel Prize for Chemistry in 1982.

Now quoting from page 78:

> Bengt Hjalmarsson (member of the Chemistry Nobel Prize Committee of the Royal Swedish Academy of Sciences):
> Science is a system ... a search driven by curiosity, all the time touching base with what's real. ... That system works. ...
> As long as they publish, someone will check their work.
> The more interesting the discovery, the more closely it will be checked. ...

Wise words; with that said, you need to be as careful as possible with all your publications and the associated data on which the publications rely.

REFERENCES

1. X. Wang, X. Ren, K. Kahen, M. A. Hahn, M. Rajeswaran, S. Maccagnano-Zacher, J. Silcox, G. E. Cragg, A. L. Efros and T. D. Krauss (2015) Retraction: Non-blinking semiconductor nanocrystals. *Nature* 527: 544.
2. C. Djerassi and R. Hoffmann (2001) *Oxygen: A Play in 2 Acts.* Wiley-VCH, Hoboken, NJ.

15 How, and When, to Effect Collaborations

It is a major possibility that at some stage of your scientific career, you will wish to take on a project of such wide scope or respond to an invitation to join such a project as a partner that collaboration is the way forward. Funding agencies also have recognised this, and there are many schemes and variants of scale of collaboration for you to choose from (e.g. see the study by Tachibana [1]).

One overriding principle to keep in sight is that the best collaboration is because of the compatibility of the (potential) collaborator with you, rather than the specific detail of what they offer scientifically.

First are your very local, intradepartmental collaborations. These are obviously convenient and can be highly rewarding. There is a huge scope for closely working with local colleagues who have highly complementary skill sets and interests. If you are looking at a new department to join, for example, as a young person choosing a faculty position, you can check out the collegiality of that department by browsing their current department members and their recent publications to see if any are collaborative. You can definitely consider this kind of collegial atmosphere a plus.

Another type of collaboration, a frequent one in my experience, is when only a few laboratories, often only two, collaborate. This may initially be on one project, and such an example I highlighted in Chapter 4, which was very successful, and led onto other ones and eventually a book with my collaborator Professor Naomi Chayen and a former PhD student, Dr Eddie Snell [2]. They are easier to manage and do not need a professional administration to run them. As Tachibana [1] states:

> Running a successful collaboration, especially one with several leaders at multiple sites, means thinking like a Chief Executive Officer: vetting partners, delegating responsibilities, and making tough management decisions. Researchers in multisite, multi-investigator projects need to adjust their grant writing approach, work culture, and even career strategy.

The perspective of funding agencies is that collaborative research is more and more the way forward:

> Around the world, funding agencies are emphasizing collaborations for many reasons – most, of course, financial. Getting experts to work together on a problem can be more effective than supporting many separate projects. Sharing equipment in core facilities avoids duplication and reduces maintenance. Many funding agencies want to see commercialization of research discoveries and translation of results into clinical practice, which requires cross-disciplinary work [1].

Governments, and the EU, also very much see the benefit of this coordinated approach not least in galvanising their research workforce to undertake research wished for by their electorates such as to tackle urgent health problems or dealing in a coordinated way with climate change, for example.

So your skills as a scientist need to be refined in the context of this major type of opportunity for research and discovery and, of course, for securing funding. And you may be the chief executive officer going about the business of vetting partners, delegating responsibilities and making those tough management decisions. Alternatively, you may be the prospective partner in a collaboration being vetted, receiving instructions, or at the rough end of tough managerial decisions. You are more than likely going to need to learn, or hone, your negotiating skills, which are formally defined as follows:

> Negotiation is the process by which parties in conflict adjust their position, by trading issues of lesser importance in exchange for issues of greater importance, because the agreement must be implemented by all parties [3].

This quotation is at the front of the handbook for the course [3] that I attended as part of my scientific civil service training as director of Synchrotron Radiation Science at the CCLRC Daresbury Laboratory. This was a highly interesting course including various business games playing out the various roles of a negotiating team. I learnt to understand my wish list thoroughly, on which points I could be flexible. I learnt to place a better estimate of value on what was wanted from me. There is also the tactical aspect of the power of a proposal. And, oh yes, very importantly, one must listen very carefully and not interrupt a proposal. I also realised how one can better recognise the exploitative type of person, which I will come back to in Chapter 31. These various things so clearly presented and practiced were highly valuable.

So where can one learn about particularly successful collaborations, as a model case study for your planned venture into your first, or next, collaboration? I offer as one such the Centre of Excellence for Coherent X-ray Science (CXS) in Australia funded by the Australian Research Council and the State Government of Victoria. This was led by Professor Keith Nugent for several years and then by Professor Leann Tilley. I was invited to join their international advisory board and served eight years on that and chaired its science advisory committee [4]. This proved to my mind to be an exemplar of why it was needed and how it was executed by all concerned. The collaboration brought together several vibrant individual research areas into a constructive whole approach with the vision to be 'the world leader in the development of coherent X-ray diffraction for imaging biological structures'. The researchers were from five academic institutions and one research institution, namely the University of Melbourne, La Trobe University, Monash University, Griffith University, Swinburne University of Technology, and the Australian Commonwealth Scientific and Industrial Research Organisation. I whole-heartedly agreed with its short summary assessment:

> CXS brings together leading Australian researchers in the fields of X-ray physics; the design and use of synchrotron radiation sources; and the preparation, manipulation and characterisation of biological samples. Regarded as a world leader in its field,

CXS aims to open a new frontier in biotechnology – the non-crystallographic structural determination of membrane proteins. CXS research is driven by its access to existing third-generation synchrotron light sources and to the Australian Synchrotron. We are also exploring the application to imaging problems of short wavelength high-harmonic generation sources and X-ray free-electron lasers that are under development worldwide.

Over those eight years, the CXS brought together annual workshops on outstanding and diverse research topics within its headline theme and brought outstanding researchers to them as speakers. The CXS researchers published a large number of high-impact across the disciplines research results that would never have happened without the CXS. The PhD student training was enriched in its diversity and again would never have happened without the CXS. The 'CXS Outreach to the Public Program' of talks and open exhibitions of the research going on and its overall objectives was frequent and rich in its efforts by all levels of staff. Finally, on a practical note, this diverse and large CXS enterprise was highly professionally managed by two dedicated administrative officers.

From my highly positive experience serving on the CXS advisory board, I also noted more generally that can-do approach in Australia that is so often said of the US research enterprise. It also seemed to be interesting, and surely not a coincidence, that the policy and the funded activity of archiving of research data, right down to the primary data, by the Australian Research Council has been world leading. See, for example, the study by Meyer et al. [5], whereby the TARDIS synchrotron data store has two missions, to facilitate mutual access to data measured at the Australian synchrotron between collaborators and to make available primary research data sets to the public. The public, lest we forget, are very much our partners (funders and potential beneficiaries) in applied and health science research enterprises. The public needs to directly see the research data on which our findings rest, and not only read our published words. The interaction with the public is our ultimate collaboration.

REFERENCES

1. C. Tachibana (2013) Science careers. Retrieved from http://sciencecareers.sciencemag .org/career_magazine/previous_issues/articles/2013_09_13/science.opms.r1300136.
2. N. E. Chayen, J. R. Helliwell and E. H. Snell (2010) *Macromolecular Crystallization and Crystal Perfection (International Union of Crystallography Monographs on Crystallography.* Oxford University Press International Union of Crystallography Monographs on Crystallography, Oxford.
3. Scotwork Negotiating Skills (n.d.) Accessed 7 August 2016. Retrieved from http://www .scotwork.com/.
4. Australian Research Council Centre of Excellence for Coherent X-ray Science (n.d.) Accessed 7 August 2016. Retrieved from http://www.coecxs.org/joomla/index.php/about -us/about-cxs.html.
5. G. R. Meyer, D. Aragao, N. J. Mudie, T. T. Caradoc-Davies, S. McGowan, P. J. Bertling, D. Groenewegen, S. M. Quenette, C. S. Bond, A. M. Buckle and S. Androulakis (2014) Operation of the Australian store: Synchrotron for macromolecular crystallography. *Acta Crystallographica Section D-Biological Crystallography* 70: 2510–2519.

16 How to Hold to a Vision Including Avoiding Politics and Carrying on Regardless of Managerialism

> Not everything that counts can be counted, and not everything that can be counted counts.

Albert Einstein

When situations get difficult involving a new development in the modern-day science work environment, I look to Albert Einstein for either inspirational words or as the test of where or how he would fit into it. The rise of managerialism in the scientific workplace is one such difficult development and is widely documented in recent books [1,2] and in regular articles in, for example, the Times Higher Educational Supplement.

Finlay MacRitchie [1] documents the rise of managerialism in the Australian CSIRO, the Commonwealth Scientific and Industrial Research Organisation, which is the federal government agency for scientific research in Australia. He describes how he, in the end, left the CSIRO and instead pursued his career in grain sciences (such as research on cereals) in the United States at Kansas State University. His book includes short biographical sketches of famous scientists in the past as inspiration to the reader and an antidote to modern managerialism, as their accomplishments did not involve being managed since they were self-driven people.

Joelle Fanghanel [2], writing more broadly than science, analyses the modern academic life via an extensive number of interviews, which included a good number of scientists. She documents how the modern academic has a range of survival methods, basically adapting to the rise of managerialism, often at a great cost, however, of the number of hours worked per week by the modern academic.

So what is *managerialism*? Wikipedia offers a definition [3]:

Managerialism is a belief in the value of professional managers and the concepts and methods they use. It is associated with hierarchy, accountability and measurement, and a belief in the importance of tightly managed organizations, as opposed to individuals, or groups that do not resemble an organization.

This approach is anathema to the university contract you will have, for example, as a professor of a given subject area, where you are the person in that university responsible for that subject area, its teaching and the contribution to developing it. As Einstein's quote at the start of this chapter neatly states, a tight management of an academic is anathema to the free-spirit and curiosity-driven scientist researcher and educationalist you need to be. In the case of the scientific civil service, in those instances at least that involve a focus on specific projects rather than research per se, its science-trained and science-oriented employees can therefore abide by a managerial approach more comfortably. Where a scientific civil service project is a facility at the cutting edge of instrumentation and methods, then freedom for its scientists to pursue ideas, and have a tolerance of reaching dead ends and of course failures, is still needed, and here tensions between staff and management can arise.

I will proceed to some of my own examples of clashes with managerialism, but first, let us analyse why a 'belief in the importance of tightly managed organisations, as opposed to individuals working independently' has arisen in the workplace for scientific research.

Harnessing science via government-funded projects to the good of society has created a focus on the progress achieved with the funds provided by the taxpayer. Taxpayers in effect then are the point of accountability for the researcher via the funding agency as its proxy. Typically, one quarter of all research grant proposals get funding. This means then that three quarters of the researchers' grant proposals are unfunded and will remain at some level only a wish in one's mind to undertake the work in some form or another. This would likely be restricted in scope for instance and over a longer time than envisaged in the original, rejected proposal. This majority of research plans, without funds but with great care and no doubt a passion of the researcher, basically explains why the researcher cannot be the subject of tight management in the workplace.

I had a head of department who, in an annual performance review, explained that I should not be spending time on research that had been rejected and was therefore unfunded. Rather I should be focusing on the next research grant proposal. From this point of view that would provide research overheads, and that is *what is counted*. I note that he did not actually stop me from doing unfunded research, which I continued to pursue through my own time and efforts and where appropriate with my final year project undergraduate students. These students would spend two full days per week with me over two semesters, and being in the top half of the class at end of the third year, required for them to be allowed to undertake a fourth year anyway, were very capable. So together, we would undertake some really nice, cutting-edge research in structural chemistry and crystallography. It may take more than one project student's work with me to reach publication standard, and over the two or three years required, we might get scooped by another research group, but at least some of the unfunded research ideas and themes I was passionate about got to move forward. I was thereby true to my university contract, as I explained it earlier, and true to my role as a free spirited researcher/curiosity-driven scientist and as a science educator bridging student training in advanced laboratory skills blended with research. However, all this was of course outside of the managerial system's wish for tight control! I should also mention that the tolerance of my efforts perhaps also stemmed from my having at any one time a good, even outstanding at times, level of research agency funds. I also made sure that I would be doing one substantive administrative role within my department at any time. I also suspect that my head of department, to be fair, was also not a little in favour of the free spirit academic himself!

Let us now analyse the one quarter of research proposals that do get funded. Overall, in the United Kingdom, there is the Haldane principle, established in the early twentieth century, whereby politicians should not interfere with what is funded, and research proposals are evaluated only by scientists [4]. A weakening of the principle has been the growth of government-directed research, namely the earmarking of funds for specific research, especially counter to the Haldane principle. So a significant fraction of the funded research is now government directed. So what about the innovative, not directed research proposals? These proposals are higher risk than average, although not always high risk, or run counter to existing science. One might then regard these as the most treasured category of curiosity-driven science proposals but, although difficult to prove categorically, tend to be rejected by the assessment panel of a funding agency. We then come to the third category, arguably the largest fraction of funded research, namely the incremental science, which basically builds safely on established science.

Overall, the progress of science, including scientific revolutions, is deemed to progress by hypothesis and falsification of existing results [5] as I described in Chapter 3. An alternative, but at times overlapping, approach is the assembly of large collections of science observational data with no clear new hypothesis in mind other than that such data sets are likely to provide an infrastructure or a framework from where new discoveries, via hypotheses, can arise. A modern-day example would be the collection of gene sequences that make up the human genome and, these days, extend to the collection and the comparison of the genomes of as many individuals as can be afforded.

In terms of which research is funded, we see a complex mixture of overt and covert managerialism tending to squeeze out innovation. Occasionally, funding agencies recognise this and set up, as the Wellcome Trust did, an adventurous research fund. Success was around 1 proposal in 30. With a collaborator in the pharmacy department, Dr Jill Barber, our proposal was one of the successful ones. With all this said, conservative, incremental research does yield significant progress. All of us can see that around us whether it be improved medicines, improvements in energy storage and production, new materials, new electronic devices and so on. We also see examples of adventurous research leading to radical change such as the jet engine or the Internet.

Should we be concerned about the low fraction of funded research proposals and, secondly, the tendency of a lack of innovation in the funded research proposals? I think we should be very concerned. So what is the answer? Such schemes as the Wellcome Trust's adventurous research scheme are important. Also, it is vital for you as a scientist to keep your lab and your own bench and analysis skills up to date because if you do not, who will carry out those unfunded innovative ideas that you will surely have? This will require the cooperation of your head of department, as I mentioned earlier, to allow you to give a fraction of your time to your own skills development and a fraction of your time to undertake your own research work. In extreme cases, as with Finlay MacRitchie [1], a job move to secure such freedoms may be necessary. In my case, I have been very fortunate not to have had to make such a move these last 25 years!

REFERENCES

1. F. MacRitchie (2011) *Scientific Research as a Career.* CRC Press, Boca Raton, FL.
2. J. Fanghanel (2011) *Being an Academic.* Taylor & Francis Ltd, Abingdon.
3. Wikipedia (n.d.) Managerialism. Accessed 7 August 2016. Retrieved from https://en .wikipedia.org/wiki/Managerialism.
4. Wikipedia (n.d.) Haldane principle. Accessed 7 August 2016. Retrieved from https:// en.wikipedia.org/wiki/Haldane_principle.
5. T. S. Kuhn (1996) *The Structure of Scientific Revolutions*, third edition. University of Chicago Press, Chicago.

Section III

---

*Being a Good Science
Research Citizen*

17 How to Referee Grant Proposals

To maintain our edge, … we've got to protect our rigorous peer review system and ensure that we only fund proposals that promise the biggest bang for taxpayer dollars… that's what's going to maintain our standards of scientific excellence for years to come.

President Barack Obama [1]

So this task is very important for us as scientists; even the president of the United States of America is taking a close interest. I am glad to say that I have always taken a great pride in this role. I have refereed for the following:

- In the United Kingdom: The Science and Engineering Research Council, the Medical Research Council, the Engineering and Physical Sciences Research Council, the Biotechnology and Biological Sciences Research Council, the Wellcome Trust and the Leverhulme Trust.
- Outside the United Kingdom: The Human Frontiers Research Programme (based in Strasbourg, France); the US National Science Foundation; the US National Institutes of Health; the US Department of Energy, the US Keck Foundation; the European Molecular Biology Organisation; the European Space Agency Microgravity Physical Sciences Research Programme; the Japanese Atomic Energy Research Institute; the Japanese Photon Factory, Tsukuba; the Canadian Innovation Awards Scheme; the Australian Research Council Discovery Awards Scheme; The Netherlands Organisation for Scientific Research; the Belgium Research Council The Research Foundation – Flanders (FWO); the Swiss National Science Foundation; the European Science Foundation Standing Committee for Physical and Engineering Sciences; the Japanese Society for the Promotion of Science; the Australian Academy of Science; The Knut and Alice Wallenberg Foundation, Stockholm, Sweden; and as an assessor for the Australian Research Council.

The two lists are approximately in year order through my career. Basically, one gets known first in your own country and then gradually around the world.

If you are asked to be a referee, you should always reply promptly to say *yes* or *no*, and if no, give your reason and provide alternative possible referees' names. To do otherwise I believe is seriously unprofessional. You would let us all, the community of scientists, down if you delay unduly in your response.

I usually find the instructions of each funding agency very clear as to what they want you to do. Their overall instructions will make clear that your role as a referee is to advise the panel, which finally holds the risks of making, or declining, an award. Basically, you will be judging a proposal for its scientific excellence, timeliness and promise, value for money and staff training potential of the project. Depending on the country of the funding agency, the United Kingdom, for example, you may also be asked to comment on such things as the proposal's possible strategic relevance, industrial and stakeholder relevance as well as its economic and social impact.

But after all, things can go wrong, and a good funding agency will have or should have a clear policy for appeals by applicants for proposals that fail. An appeal might be deemed valid by the agency if the applicant describes evidence of factual errors, bias or conflict of interest of the referee or evidence of a lack of expertise in the judging panel. Obviously, an appeal cannot be based on opinion alone. An appeal would have to be evidenced based. So, a referee's report needs to be carefully done and factually accurate!

Overall, I hope that it is obvious that refereeing is a skill, and you should make your expertise in science available in a timely and appropriate manner. Also, be constructive. You will be on the receiving end of referees' reports too of course with your own proposals. So do unto others as they would do unto you. But also, do your duty by the taxpayers who ultimately fund the research including you!

It is also worth highlighting that refereeing research grants is one of the requirements for career advancement, as judged by your department. It is also hugely helpful in writing one's own grants!

REFERENCE

1. National Institutes of Health (2013) Peer Review Process. Accessed 7 August 2016. Retrieved from http://grants.nih.gov/grants/peer_review_process.htm.

18 How to Referee Science Articles

Over the 40 years or so of my research career, I have refereed for the following journals:

- The International Union of Crystallography's (IUCr's) *Acta Crystallographica, Journal of Applied Crystallography, Journal of Synchrotron Radiation* and *International Union of Crystallography Journal*
- *Journal of Molecular Biology*
- *Biochemistry*
- *Biochemical Journal*
- *Biochimica Biophysica Acta*
- The Royal Society of Chemistry's *Faraday Transactions, Laboratory on a Chip, Physical Chemistry Chemical Physics* and *Dalton Transactions*
- The American Chemical Society's *Journal of Physical Chemistry* and *Journal of the American Chemical Society*
- *Crystal Growth and Design*
- The American Institute of Physics' *Physical Review Letters, Physical Review B* and *Review of Scientific Instruments*
- The Institute of Physics' *Journal of Condensed Matter Physics* and *New Journal of Physics*
- The Federation of European Biochemical Society's *Letters*
- *Environmental Biology of Fishes*
- *Bioorganic & Medicinal Chemistry Letters*
- *European Journal of Biophysics*
- *Comparative Biochemistry and Physiology*
- *Molecular Biology and Evolution*
- *Nature*
- *Proteins and Peptide Letters*
- *Science*
- *Current Opinion in Structural Biology*
- *Chemical Physics*
- *Coordination Chemistry Reviews*
- *Structural Dynamics*

This diverse collection of journals in the list above partly reflects not only my broad interests but also as a specialist in crystal structure analysis using X-rays and neutrons, which results can impinge across many science disciplines. One of the Committee on Publication Ethics' (COPE) rules for a referee [1] immediately kicks in, namely that upon accepting an invitation to referee, you need to explain to the editor what portion of the submitted article you are professionally qualified to comment on. The COPE rules are an excellent guide to pretty much all the issues that may arise for you, and you should adopt them as your skills base of how to approach the task properly. COPE is a forum for editors and publishers of peer-reviewed journals to discuss all aspects of publication ethics. It also advises editors on how to handle cases of research and publication misconduct. While the president of the United States has not personally taken an interest in the importance of the role of a referee for a journal, unlike as in the last chapter, the role of the referee for grant proposals, I can assure you, it is a very important role.

It is vital for you to read the specific journal's instructions. These can vary quite a lot most significantly on whether the journal wants you to put a numerical value on the likely impact of a paper, for example, 1–5. I find this is a very difficult question. Specifically, how would someone know how well a paper is going to be cited?! I also find it rather distasteful, but I do respect the wishes of the journal and make my estimate. Why do I dislike that question? A journal will never say how many years it might take for the article in question to make its impact. This is a quantity that is measured formally by the citation half-life, which quantifies, more or less, when citations to a paper stop. This can also vary a lot between fields. Roughly speaking, biology and biochemistry papers are supposed to have a shorter citation half-life than other scientific fields. Citations can also occur for the wrong reasons such as being in error; a famous, high-profile example was the cold fusion as an energy source report.

Another aspect of the notes for referees of a given journal is whether they mention the availability of the data on which the words of the paper are based. I find that the availability of the data is unfortunately very patchy. Increasingly, I am asking for the data as a referee, and I will examine these closely. When an author refuses my request for the data, transmitted via the journal editor obviously, the process of my evaluation of the submitted article has stalled, and I can offer no report. If the editor has any sense, and they usually do (but not always), the paper is rejected.

A major trend of all journals is the demand of the referee for speed to get their report back. This to me is a false logic on their part. At its heart, a journal article, compared with a newsletter article, has a long-term archive role. What is three to four weeks versus 10 days for the refereeing that is so important to the authors who press journal publishers to provide fast turnarounds? That is simply madness, as, in my view, authors are protected against their research competitors, where such exists, by the article submission date. Anyway, when I get a request to reply in 10 days, and am only provided an abstract to make my decision, I will request the manuscript to properly decide if I can provide a proper report in the time required. What is a proper report? It is one you can defend when the authors write their replies (rebuttal replies).

There are various other points, and I refer you simply to a very good, cogent statement of the roles expected of a referee for the journals of which I was editor-in-chief between 1996 and 2005, the *IUCr Journals* [2].

So what will most likely be your decision on a paper you have agreed to referee? I suggest that this will be between the extremes of the following:

> It is always a joy to review manuscripts such as this. Well-conceived, well executed, well edited. Clean. Pristine from start to finish. Accept as is.

And

A weak paper, poor experimental design, no statistical analysis possible, carelessly written, poorly thought through. Reject.

These two overall summaries are from *Chemistry Blog* [3]. I have added my own final decision words. *Chemistry Blog* [3] also offers, as well as those two serious statements, quite a list of humorous extracts from the referees' reports that the *Environmental Microbiology* journal's editors have received. These humorous quotes were highlighted on BBC Radio 4, a popular channel in the UK [4]!

You will spend the majority of your time deciding between major revision and minor revision recommendations, and subdividing a bit further, mainly deciding between major minor revisions or minor major revisions.

All the decisions of course need to be followed by a careful and specific set of constructive details of changes required. Again, do unto others as you would wish to be treated yourself. Also, you need to think very carefully at the 'submit your report?' moment, rather than simply think 'job done'. After all, there are examples like this:

Hans Krebs' classic paper on the Krebs cycle, perhaps the most important development in cell biology in the 20th century, was rejected by *Nature*. They said it was of insufficient general interest. He went on to win the Nobel Prize for that [4].

REFERENCES

1. I. Hames (2013) COPE ethical guidelines for peer reviewers. Accessed 7 August 2016. Retrieved from http://publicationethics.org/files/Ethical_guidelines_for_peer_reviewers_0.pdf.
2. *Acta Crystallographica Section E* (n.d.) Accessed 7 August 2016. Retrieved from http://journals.iucr.org/e/services/referees/notes.html.
3. *Chemistry Blog* (2010) Accessed 7 August 2016. Retrieved from http://www.chemistry-blog.com/2010/12/18/referees-quotes/.
4. T. Feilden (2010) Scientists have sense of humour, shock. Accessed 7 August 2016. Retrieved from http://www.bbc.co.uk/blogs/today/tomfeilden/2010/12/scientists_have_sense_of_humou.html.

19 How to Write a Balanced Book Review

During my career I have written about 20 book reviews. As a joint main editor of *Crystallography Reviews*, and finally main editor, I have commissioned up to about 100 book reviews. As the recipient of an invitation to review a book, I have always regarded it favourably, basically as a commendation to write an opinion on a book within my area of expertise. It is also one of the most challenging of all commissions to take on; as a reviewer, you balance the feelings of the author and exercise your duty to the potential buyers of the book. As a book author myself, I know that state of anticipation, and excitement, as you await the reviews of your book. It is then a considerable skill to write a fair and insightful book review, which also needs to be interesting in its own right. For the latter, you have licence to add a few details of your own, in the hope to make a wider context for instance.

So how do you start approaching this task, which I think will surely come your way as an active researcher scientist and as you build up your research reputation? Clear instructions from the person sending you the invitation are essential. Journals vary in the length of a book review that they expect; they can be as short as 500 words and as long as 1,500 words. There are also useful guidelines like 'State the author's aims in writing the book and then scrutinise if they have succeeded in those aims'. Also, 'Are the aims of the author good aims?' On occasion, as a commissioning book review editor, I have forgotten to include that in my reply when a person has accepted, or simply thought that such basic instructions would surely not be needed for the very experienced person I had invited. But no, when submitted, I could see how the person had not perceived the need to explain that at all.

It is good also to look up reviews of books that you know. You will find these in the journals that publish your research. You can then compare the different styles of book review. Where a book deserves criticism, the styles of how to point out the errors and deficiencies can vary a lot. One common aspect that is however essential in such cases is for you as the reviewer of the book to be specific and evidence based with your criticisms.

The reader of your book review also needs a clear final recommendation. If you recommend purchase of the book then you need to say if it would be for a library, a laboratory's shared collection of books or for someone's personal office. There are also the different types of version now: hardback print, paperback print and e-book (for Kindle or iPad, for example). Also, when writing your final recommendation to buy, you would consider whether the new book is sufficiently different to previous books covering the research field and the price of the book as to its value for money.

So of my various book reviews, which have I enjoyed writing the most? Also, of the book reviews I have commissioned, which have I enjoyed reading the most? More widely, which are my favourite book reviews? I will start with the latter.

A book reviewer who is a Nobel Prize winner reviewing the book authored by another Nobel Prize winner to my mind clearly has potential to be a good read. My overall favourite is Max Perutz's review in the New York Review of Books of Peter Medawar's *Advice to a Young Scientist*. This book review has been reproduced in full in Max Perutz's book *Is Science Necessary* [1]. The most read book review in *Crystallography Reviews*, combining much humour within the careful review of the material, downloaded almost 1,000 times, is that of Professor David Rankin, a specialist of gas phase electron diffraction at Edinburgh University, of the *Handbook of Chemistry and Physics* 89th edition [2]. The review of a book on the incredibly diverse topic of research ethics was taken on by emeritus professor of physics at Keele University, Watson Fuller, a superbly insightful coverage of the moral aspects of scientific and medical research [3]. Another reviewer, Dr Joe Ferrara, clearly enjoys reviewing books on many different subjects; for example, he accepted my request to review the autobiography of an outstanding scientist of our field, Francis Crick, with the unusual feature of the book being published many years before [4]! Another very careful book reviewer is Professor Massimo Nespolo from Nancy University, France, and now is himself a book reviews commissioning editor (for IUCr Journals); see, for example, his book review [5].

Of my own book reviews, I will allow myself to mention two. My book review of *What a Time I Am Having: Selected Letters of Max Perutz* [6], A collection brought together by his daughter, Vivien Perutz. In the book review of *Early Days of X-ray Crystallography* [7], I reviewed the book by an IUCr past president. Andre Authier. The seniority of these two people presented a great challenge to my skills and experience with book reviews. I shared my own insights nevertheless; for example, in my book review of Max Perutz's letters, I stated the following:

> Much later in this book I suffered another challenge to my crystallographic beliefs when I came upon the letter written in New York on 1 May 1995 (pp. 467–468) to his son Robin, whom my wife and I know quite well, where Max states 'I did not find myself shedding many tears for my US colleagues' shortage of funds. As usual on my visits, most of them have just 'happened' to move into new buildings or collected superb new data at Grenoble, or replaced their almost new image plates by a new superior charge something device. Crowds of them are tearing the guts out of structural biology, and most protein structures now emerging resemble one or another seen before'. Whilst structural genomics, to which Max is presumably referring, has excited wide views in the field, pro and con, I was however shocked at his adverse comments on the pace of technological change, which has surely greatly enhanced all aspects of the field of structural molecular biology, and at relatively modest cost.

I was bound to pick up on this as I had a great deal to do with the establishment of those facilities for structural biology at the European Synchrotron Radiation Facility (ESRF) in Grenoble, not least as chair of its European Working Group for the foundation phase of the ESRF in the 1980s, and later as vice chairman and then chairman of the ESRF Science Advisory Committee. I believe I successfully advised future readers in an insightful and balanced way. I concluded my book review as follows:

This book is much more than the few items I have highlighted, which were the letters and quotations that particularly challenged me, and which therefore served as examples of the deep interest one can take in these writings. But I am not going to reveal whom I label 'X' in the sentence: '(I) sympathize with the view that person X is a charlatan'; you will just have to buy this book to find out. Likewise the chapter of last letters (pp. 483–494) is a very moving collection in its own right. It is a privilege to have access to all these letters in this remarkable and splendid volume.

REFERENCES

1. M. Perutz (1991) Book review of *Advice to a Young Scientist* by P. Medawar, Sloan Foundation Series 1979. In *Is Science Necessary?* Oxford University Press, Oxford: p. 196.
2. D. W. H. Rankin (2009) Book review of the *CRC Handbook of Chemistry and Physics 89th Edition*, D. R. Lide (ed.), CRC Press, Boca Raton, FL. *Crystallography Reviews* 15(3): 223–224.
3. W. Fuller (2009, October) Book review of *Research Ethics* (2008, September 4), K. D. Pimple (ed.), Ashgate Publishing Hampshire, England and Burlington, VT. *Crystallography Reviews* 15(4): 289–293.
4. J. D. Ferrara (2012, October) Book review of *What Mad Pursuit: A Personal View of Scientific Discovery* by F. Crick. *Crystallography Reviews* 18(4): 308–311.
5. M. Nespolo (2016) Book review of *X-Ray Crystallography* by G. S. Girolami (2015), University Science Books, Mill Valley, CA. *Acta Crystallographica* A72: 168–170.
6. J. R. Helliwell (2011) Book review of *What a Time I Am Having: Selected Letters of Max Perutz* (2009), V. Perutz (ed.), Cold Spring Harbor Laboratory Press, New York *Crystallography Reviews* 17: 2, 151–152.
7. J. R. Helliwell (2014) Book review of *Early Days of X-ray Crystallography* by A. Authier (2013), International Union of Crystallography/Oxford University Press, Oxford: pp. xiv, 441. *Acta Crystallographica* A70: 92–94.

20 How to Be a Science Research Editor

Editors bear the full responsibility to ensure that journal content is ethical, accurate and relevant to the readership. They work closely with referees within the peer review framework to achieve this. A comprehensive description of peer review and manuscript management in scientific journals, including its golden rules, can be found in the work by Hames [1]; this book also has example, template, letters to authors, including in various circumstances, and referee reply forms for many different journals are described.

My work as a science research editor involved 25 years of devoted service to my learned society publisher from 1990 to 2014. In the 9 years between 1996 and 2005, I was its editor-in-chief and chairman of its journals commission. I was also responsible for the specific science that was either submitted to me as research articles, about a thousand in number, or for chairing the scientific and sometimes policy discussions at the journals commission. I worked closely, especially as editor-in-chief, with the managing editor, and I represented the journals at my learned society's finance committee. The journals I was involved with were those of the IUCr, namely *Acta Crystallographica Sections A* through *F*, *Journal of Applied Crystallography* and *Journal of Synchrotron Radiation*.

The overall vision of a learned society publisher at its core is to provide that research community with a place to publish their work free of commercial imperatives and possible interference. Secondly, if at all possible, the aim is to make a significant enough surplus to support the sustainability of the journals' operation and to support the good works of the community through student bursaries to attend conferences and other similar activities as well as promotion of crystallography as a discipline via outreach activities. A big outreach activity was the United Nations Educational, Scientific and Cultural Organization-endorsed International Year of Crystallography in 2014 led by the IUCr. Thirdly, through meetings of all editors and coeditors, broad discussions, from all corners of a research community, can come forward. In my nine years as editor-in-chief, the IUCr journals commission meetings, held every three years, comprised well over a hundred scientists from many research disciplines. These community meetings would be held just before an international congress to keep attendance costs to a minimum. These meetings sustained the vision of the learned society as a cooperative, everyone working together. We would discuss together best practice, journal updates, technology developments and possible future direction and development of the publications. I think this cooperative discussion approach is very important in the learned society context. We also responded to community inputs and wishes, via open discussions at major conferences, with the launch of new initiatives. The role of editor-in-chief, and indeed all the existing coeditors, included attracting good scientists to the scientific editorial team, but with appointments finally made by the IUCr executive committee, an elected body of the IUCr general assembly. This body comprises almost 100 delegates from all the adhering countries of the IUCr, which is the 'United Nations' of crystallography active countries in the world! The important point here about the governance of the IUCr is that the IUCr executive committee is independent of the IUCr journals commission.

As an editor-in-chief, I decided that I needed to fully understand the libraries' points of view of science journals. So I decided to volunteer to serve on the University of Manchester's Faculty of Science Library Committee. I duly became its chairman and then represented the faculty at the University Main Library committee. There I interacted closely not only with the librarian with the responsibility for science books and periodicals but also the main librarian. I already knew that science journals were a costly subscription operation for the University of Manchester, and now I got to learn the hard facts that were involved, namely the very large profit margins that some commercial publishers can make (reported to be around 40%) from the science articles whose research content was funded in considerable measure by public money. I also found out that not all learned society publishers made 'surpluses for the sustainability of their journals and the rest to support education and training of needy scientists within its ranks as well as outreach to the Public'. Some learned society publishers, my main librarian told me, also made large profits. Librarians had tried collective deals organised by country. But still, the journal subscriptions kept on going up. An important initiative in this period was the Budapest Open Access Initiative [2]. Ultimately, the government and some charity funding agencies decided that enough was enough, and an important practical step was taken of gold open access, with articles paid by the science authors themselves to publish their research. This put the responsibility of what article processing charges were fair and reasonable directly in the hands of the scientists themselves and can work well for funded research (for unfunded research, see Chapter 7). It always amazed me how some other learned societies, who my main librarian lumped together with those high-profit commercial publishers, did not do as good a job at serving their research communities as ours did, the IUCr.

Overall then, I had two reasons to give up some of my spare time to undertake the role of scientific editor, as I was already very busy as a university professor. The first reason was that I felt, most simply, that as a scientist myself, I had received help to improve my own research articles via the peer review organised by my society journals, and so by being an editor myself, I could be of assistance for others in return. Secondly, my efforts would help create some revenue to support our crystallography community students' training and so on.

It was a bonus that being in an international society, I could meet scientists from all over the world and occasionally support them to get their articles up to publication standard. By this, I do not mean help them with their English syntax, which was not my primary role, but by sharing my science expertise. With some authors, whether from the developing or the developed world, I became conscious that not only the referees' reports but also my own comments to the authors could help the authors towards improved science through further experiments and or analyses they really should undertake. The helpful and constructive approach that I took was not always compatible with a high-impact factor philosophy. Also of course, constructive help such as me advising further analyses or experiments would contradict the aim of speedy publication, i.e. very short times between submission and acceptance dates. The authors of course want prompt editorial handling if possible but, in my experience, do understand constructive suggestions. A specific yet important difference between 'further analyses being needed' and 'further experiments being needed' of course is that while for the former, I would preserve the article as a submission with its date, but I could not do that for the latter, which would require the reply 'rejection with invite to resubmit' since new data were needed.

So from what I have described earlier, science expertise is at the core of your role as a scientific editor, if you take on the role. It is a very demanding role. You gain help and support from the referees that you choose to consult for a given article submission; either I would always ensure that the referees were happy with a revised version or if I disagreed with some aspect of their report, I would always write and explain why. But finally, the acceptance of an article was my decision. My final read of an article was extremely important in that balancing act in the role of being fair to the authors, the IUCr's journals as publisher, the journal subscribers, our readers, the funding agencies and, increasingly more obvious, the public. Further revisions of an article at my final read stage were always possible, but a rejection at that late stage would be rare, as the referees would surely have got to that point ahead of me.

Overall, this due care and attention to process, and being as constructive as possible to the authors' and their research and everyone I have listed earlier, takes time. If you are not in tune with the journal's vision and ethic, who initially invited you, do not take on the role with them; instead, await the invitation from another journal. Importantly, I encourage you to try and help your learned society if at all possible; that is what I have done during those 25 years, and associated with that role, I judged about 1000 science research articles. I hear you ask what was my article rejection rate finally? Answer: Around 25%. I could also see that my constructive approach provided a solid science literature as well, having gone closely through the research work with the authors. All the articles I accepted as an editor have done well; not one was retracted, and the journal impact factor, which I do monitor but I am not a slave to, always held up well compared with other science journals. My application of the criteria as to whether the research reported in an article submission were significant and new, and that it had real findings and actual conclusions, held up well basically.

After I retired from being the editor-in-chief of IUCr Journals, I continued to serve as a coeditor for the *Journal of Applied Crystallography* for a further nine years. The IUCr asked me to serve subsequently as its representative to the International Council of Scientific and Technical Information (ICSTI) [3], which I did from 2005 to 2014. ICSTI is a fascinating and important forum for sharing many perspectives, technical developments and policy on these vital topics for scientists, librarians and providers of such information. On my retirement from that role, after the usual IUCr's nine-year service guideline, I commenced the role in 2014 of representing IUCr at the International Council of Science's CODATA (see CODATA [4]). CODATA works to improve the quality, reliability, management and accessibility of data of importance to all fields of science and technology. As part of this role, I am chairing the IUCr's Diffraction Data Deposition Working Group looking at all the aspects of extending data access to include raw diffraction data, i.e. before data processing and interpretation; see the study by IUCr [5].

In addition, since by now I had gained a very wide perspective of crystallography, I approached the main editor of *Crystallography Reviews*, Professor Moreton Moore of Royal Holloway College, London University, if I might join him as editor. This did not interfere with IUCr Journals as they do not publish full reviews i.e. comprising around 20–30,000 words in length. Moreton and the publisher, Taylor & Francis, agreed. The role of main editor on this reviews-only journal has the major difference of the review articles being largely commissioned by the editors or by the members of the editorial board, although some are naturally submitted at an author's initiative. Taylor & Francis also publish many journal titles across sciences, social sciences and humanities, including one of the oldest science journals *Philosophical Magazine*, founded in 1798. I have attended three of the Taylor & Francis Editors' Round Tables; two events were held at The Lowry in Salford and one in Leeds. At these events, I learnt details of a great diversity of journals and discussed old and new issues such as peer review challenges, open access, changes in technology, use of social media etc. At the core, this is a commercial company, not a cooperative learned society (although individual learned societies have their house journals published by Taylor & Francis). The diversity of academic subjects within this commercial enterprise showed me that there are benefits of the spirit of private enterprise to scientists, as long as the company concerned keeps their profits at a reasonable margin!

Overall, on looking back on my scientific editorial career, it is a great feeling of having ensured that the journals I have been responsible for had content that is accurate and relevant to their readership and of course ethical (I return to the topic of ethics in Chapter 31). I have also managed through my service to support the global crystallographic community in numerous educational activities. So why not volunteer yourself!

REFERENCES

1. I. Hames (2007) *Peer Review and Manuscript Management in Scientific Journals: Guidelines for Good Practice.* Wiley-Blackwell, Malden, MA.
2. Budapest Open Access Initiative (2001) Retrieved from http://www.budapestopen accessinitiative.org/.
3. International Council of Scientific and Technical Information (1984). Accessed 7 August 2016. Retrieved from http://www.icsti.org/.
4. CODATA (1966) Accessed 7 August 2016. Retrieved from http://www.codata.org/.
5. IUCr Forums (2011) Public input on diffraction data deposition. Accessed 7 August 2016. Retrieved from http://forums.iucr.org/viewforum.php?f=21&sid=28ceb2dfe5bb4e1c89 2110256fd86447.

21 How to Chair Meetings

A committee (or 'commission') is a body of one or more persons that is subordinate to a deliberative assembly. Usually, the assembly sends matters into a committee as a way to explore them more fully than would be possible if the assembly itself were considering them. Committees may have different functions and the type of work that each committee does would depend on the type of organization and its needs. [1]

Meetings come in all shapes and sizes! Quite often conferences, even very large ones, are called *meetings*. However, let us focus here on committee meetings.

The most difficult meeting I ever chaired was a joint meeting of the ESRF Machine Advisory Committee (MAC) with the Science Advisory Committee (SAC) held in Grenoble in 1992. This meeting of around 40 specialists of accelerator physics and others from across many science research disciplines and techniques using light, as well as the management of the ESRF, had to deal with the very difficult issue brought about by a contractor incorrectly laying the concrete of the ESRF experimental hall floor. The meeting brought the issue to a head with all competing factions and personalities present. Two extreme views were expressed. The director general, Professor Ruprecht Haensel, emphasised a civil engineering solution known as *shallow grouting*, which involved squirting liquid grouting compound between cracks in the floor slabs every 12 months; this seemed immediately unsatisfactory, as it was meant to be a clean scientific working environment, which then would be forced to be a building site every 12 months. The contrary view championed by a leading member of the SAC, Professor Michael Hart, was that the concrete floor should be dug up and taken away by the contractor who should then lay a new floor; this option seemed also unworkable as not least the sheer quantity of concrete to be dug up and lorried away was very large. Dr Massimo Altarelli, from within the ESRF management, proposed that the civil engineering consultants continue to hunt for a more long-lasting solution to the problem. This last option, which faced the joint MAC/SAC meeting, had the danger of looking like procrastination if we took it! But in spite of that disadvantage, it was the only realistic thing to commend to the ESRF council, comprising all the formal representatives of the country partners, 11 in all at that time, and who had the ultimate authority over such decisions for the ESRF. As ESRF SAC chairman, I attended the next ESRF council and had to convey the details of the outcome of the joint MAC/SAC meeting. Professor Jules Horowitz was in the chair and was greatly experienced in all manner of large building projects. The civil engineers had by that time offered a deep grouting solution to the problem and for which they estimated that the concrete floor would be rendered stable and sufficiently flat for up to 20 years. This was received with great relief by all delegates at the ESRF council, not least as the engineers' safety factors in any project would likely be such that it could be good up to twice that length of time! While the whole problem was lamentable that it had even occurred in the first place, it was a profoundly satisfying business to see it resolved. Of course, most importantly, it was and is a great thing to be involved with such a supranational pan-European project.

I start this chapter with this example because it illustrates all sorts of aspects of the importance of committee work on the one hand and the personal as well as scientific skills that will be needed of you as a member of such important committees and, in due time, when you may well be the chairperson of such important committees. I use the words *important committees* in the previous sentence. After all, you are a busy person, and you will need to prioritise what responsibilities you take on. If you are interested, however, then you can observe such persons as you admire to see just what committee tasks they take on. Different people have different domestic responsibilities and different energies as well as temperaments, but overall, at any one time, I think it good that you would take on one at least important committee role. You will have to make some adjustment for what you regard as important according to your level of experience; you cannot expect to be chairperson of the most important committee in your field just like that. Also, to start with, you will be a member and not the chairperson.

Having taken on the role of chairperson, you will have previously studied the good practice of your current chairperson or the bad practice if you have been unlucky enough to have been serving under a poor leadership. What then are the skills of a good chairperson?

My starting example involving a science advisory committee of the ESRF had obvious and broad terms of reference, namely all science connected to the ESRF. A committee can also be established with a very specific aim in mind. One such was when I was asked by the IUCr to chair a diffraction data deposition working group. A specific set of terms of reference keeps such a committee focused, and avoids mission creep. This latter term is, I imagine, from military circles and where mission creep could prove especially disastrous; anyway, it is certainly a very evocative term.

There are some very practical aspects as the chairperson of a group of people in a committee that you should remember. Also, there are many types of committee [1], and so I will outline a procedure for one type, namely where a host organisation has brought you together as an advisory committee with an agenda that has been agreed with you well before the meeting and circulated to the members for comment. Each agenda item will be accompanied by briefing details and questions to the committee. I think the specific details I give will assist you, however, with many types of committee in its individual parts at least, if not all the steps I now list.

Firstly, at the start of the meeting, you have the chance to create a collegial, collaborative tone to the discussions by asking each person in turn to say who they are, where they are from and, where relevant, who they might be representing. This is known as the *tour de table*. To be purely arbitrary about it, I start from my left and go to each person in turn around the table. I usually explain my details at the start, as an example of the detail required.

Secondly, since you have carefully already arranged with the hosts specifically timed morning coffee and afternoon tea breaks, you should keep to the times set for these, not least as the hot drinks can cool off and become a disappointment to the members rather than a nice refuelling. As Napoleon is attributed to have said,

'An army marches on its stomach', meaning a well-fed and watered group of people will be in the right frame of mind for tackling the, almost certainly, challenging issues the committee has been brought together to address. Yes, water should also be available throughout the meeting; a thirsty committee member is not going to be in a good frame of mind for long!

Thirdly, there may be a presentation by a member of your host organisation to set the scene for that agenda item. As chairperson, you can open the questions or invite a member of your committee, most well versed in that item, to start the questions. Where there is no presentation, you can introduce an agenda item with a very short summary, as the meeting already has the relevant background information. You then, mainly, listen carefully to the members' inputs and then make your first attempt to provide a summary. Before your summary, you have to remember who has not yet spoken and invite them to make their comments known, or make a more general query for anyone else to make an input if the committee is large.

Fourthly, on each agenda item you must finish that item by clearly stating what has been agreed and ensure that your committee secretary has had the time to write that down, or these days, type those recommendations into a MS Word document yourself.

Fifthly, you will probably need to present your recommendations in a short close out presentation, and as a group, you will have a final time slot to prepare together, with assignments to each committee member, or as a pair, a couple of slides for each topic. You as chairperson would assemble the slides together and then give the findings of the committee to your hosts.

Since this type of meeting involves a lot of discussion with your hosts, you need blocks of time set aside for your own, confidential, discussions in closed sessions.

Also, after day 1, I would try and get everyone relaxed with some fun speech making at the SAC dinner.

You and your committee secretary will be responsible for bringing together the meeting summary. The individual sections of this summary can be most readily compiled from your individual members, each writing a section for which they are most knowledgeable, as with the preparing of the slides for the close out presentation mentioned earlier. Recommendations have to be worded particularly carefully and be clear. These will be put in bold typeface.

My own style and attention to detail on this role was pretty good I thought, as I have chaired many committees, but the director of the Spanish Synchrotron Radiation Facility (ALBA), Dr Caterina Biscari, sharpened my approach further, embodied in the above notes, and I record my gratitude to her here for that. I am also very grateful to Professor Joan Bordas, as the ALBA's founding director, for first of all inviting me to join the ALBA SAC; I undertook the role as its SAC chairman between 2010 and 2014, and also was president of the ALBA Beamtime Application Panel. On my retirement, I received a collection of gifts among which was a T-shirt for me and one for my grandson, with a short, indeed beautiful message (Figure 21.1).

In the earlier description, please note that I carefully made the distinction between *recommendation* and *decision*, the point being that your committee is advisory, as I explained initially.

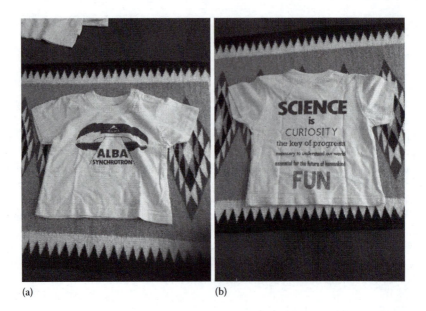

(a)   (b)

FIGURE 21.1   An important T-shirt message from the Spanish Synchrotron ALBA! The two pictures depict the (a) front and the (b) back of the T-shirt.

One other aspect of a committee of individuals is the possibility that one or more members may have a conflict of interest. When I was the chair of the School of Chemistry Examiners' Meeting, as chief examiner, which I undertook for four years or so, I was obliged by the university's regulations to open the meeting with a formal statement that anyone present with a relationship to a student should declare that. As anonymous marking was introduced and all students became a student identity number, this statement was basically rendered obsolescent. Other situations where conflicts of interest have to be dealt with are committees considering research grant or beam time applications. An application coming from the same institution as a member of the committee would mean that the committee member should leave the room. In the case where the chairperson had to declare a conflict of interest, it would of course mean that someone else should chair that particular discussion. Such a situation would also present you with a good occasion to practice your skills as a chairperson if you are the one to step in.

REFERENCE

1. Wikipedia (n.d.) Committee. Accessed 7 August 2016. Retrieved from https://en.wiki pedia.org/wiki/Committee.

Section IV

Skills for Being an Educator

22 How to Teach Your Subject to Undergraduates

We saw in Chapter 6 the example of the teaching that Sir Peter Day took so much exception to in his autobiography [1], and my reaction to it in my book review [2]. Also in Chapter 6, I introduced the details of the New Academics Programme (NAP), with an emphasis on your training for being a lab leader. The NAP naturally offers a wide range of training in options for teaching. A key feature of this is the teaching observation form, which requires an assessment of the following aspects of your lecture:

- Planning and preparation
- Management, communication and interaction with your audience
- Resources (which hardware medium) used
- Finally, various comments given by the observer, an experienced academic

Of course, a lecture is a communication. There are some very basic points to consider! If the student cannot hear the lecturer or read the screen/blackboard, then the communication has failed. Student questionnaires should ask the fundamental questions: Could you hear the lecturer? Could you read the screen? – (rather than asking, 'Did you have a quality experience?').

I think we can readily imagine how Sir Peter Day would have wanted his lecturer, Dorothy Hodgkin, to be assessed! But that would be his views to express on his student questionnaire form. It would be difficult to imagine much support by the observer 'of a lecturer just turning up with a few of their research slides'.

The National Student Survey in the United Kingdom, introduced in the 2000s, is clearly interested in student satisfaction, and indeed, it is the student who fills in his or her form. Separately, employers make their views known informally, for example, via the media with headlines like graduates have/do not have the skills we need. Academics themselves and scientists in particular have a voice, albeit indirectly, via their professional societies such as the Royal Society of Chemistry and the Institute of Physics, who give external accreditation to degree courses. Then we have of course examination degree results and degree classification outcomes now increasingly supplemented, although not yet replaced, by marks transcripts. I served four years as chief examiner in the School of Chemistry at the University of Manchester and thereby learnt the myriad of details that can arise as special circumstances for mitigation on behalf of a student where there was evidence, for example, a doctor's certificate. Overall then, there is a complicated ecology, and it is an ecology, within which you will seek to educate and indeed inspire your students about a subject. This requires above all else a skill known simply as *your philosophy*. You can have a profound influence on your students. You will receive thank you cards of gratitude from your final-year project students and occasionally applause at the end of a lecture course. As an enthusiastic presenter of lecture demonstrations [3], I also received applause during a lecture after a laser light demonstration of diffraction from a two dimensional grating, which I had then spun around with a motor at an increasing speed until it gave a powder diffraction pattern. But you will also receive unappreciative comments via the anonymous student questionnaire from students who find your material difficult or where you perhaps went too far in recommending too much further reading.

I have taught physicists, chemists and biochemists/molecular biologists. The experience of the students in maths greatly varied in these subject groups, and at times in my teaching, I needed to use maths. I tried all permutations and eventually found that I would need to teach those parts of a course in two different ways, with the maths and without the maths (hence, for the latter, the utility as well as the diversionary attractiveness of lecture demonstrations!). This worked well, but examination question setting needed special care. The first year that you give such a course requires providing students with typical examination questions with answers.

Throughout my career, although it has perhaps increased in recent times, there would always be the question from students 'will this be in the exam?', usually when a more difficult portion of the lecture material was covered. This is a key moment as a lecturer. I would answer, 'Anything I have covered in the lectures might be in the examination but if you find a topic difficult then this is where the tutorials/ workshops will surely help you understand, and you also have the Library'. I could also realistically offer to provide special tutorials myself alone for subsets of the class, for example, 4 or 5 students at a time. An interesting development, useful for larger classes, for example, 200 students, is the chat room within the university's electronic blackboard system. With this, I could monitor a discussion as members of a class who did understand something would answer queries for others on the course. I would step in if a discussion thread went wrong or in an unhelpful direction.

One aspect that has evolved fairly constantly, with the development of technology these last four decades, is how you present your lectures. However, some things never change! First, your undergraduate lectures are not the same as your conference research results lectures. The only exception to this rule is in the final year of the degree course when you will interweave recent research results or unsolved research challenges into your lectures. There are some common aspects to get right in both types of lectures, conference research or teaching lectures. Whichever technology medium you use, it has always been the case that you should do the following:

- Never present anything that you do not, or even worse, cannot explain.
- Never teach anything in a rush.
- Never teach anything too slowly.
- If you realise that there was a mistake in your material, then at the earliest opportunity, make clear that it was in error and give the correct material.

There are other rules that have developed with technology. So with PowerPoint material, never assemble too many slides for the time allocation; this is easy to do in advance but to be absolutely avoided. Also, never load a slide with too much material. In the old days with chalk and a blackboard, such hazards never arose, although the lecturers who would try and speak at 200 words a minute were a nuisance, speaking I mean as a student myself at one time.

New opportunities have arisen from technology developments. Firstly, a material can be placed on the teaching blackboard ahead of the lecture, but you can set the release date after your lecture, which is important to encourage learning by the students' writing it down themselves. Secondly, however, printed handouts given to students at the start of a lecture became the norm, not the exception. So how to facilitate learning and students' attendance!? One strategy for the lecturer is to leave gaps in the material, and the student has to concentrate sufficiently to fill in those gaps. This method has perils of its own, namely trivial or near trivial material for those gaps. Another, more recent approach, is to blend the PowerPoint projection with projection via overhead transparencies on a second screen. These transparencies can be written on at the time, which is better, or be preprepared. The use of a digitiser can capture the digital image of the transparencies. A quite contentious topic has been the introduction of the full recording of teaching lectures; this initiative is quite recent and unclear on its likely success and impact on examination results. In my experience, the most successful strategy did vary for the year of the class and the size of the class and whether the topic was quite specialised or quite general. This is the fun of the teacher, that of trying out different approaches, and indeed a key element is that every teacher has their own style and personality and thereby preferences. It is important to be reasonably fluent. Everyone I believe gets nervous before and at the start of any lecture, and so you will work out your own way of achieving fluency and overcoming any nerves.

One unexpected thing happened to me that was very interesting. I was assigned a course at a fairly short notice (a couple of months). The previous lecturer was very companionable and indeed professional and provided me with all their teaching materials. What was interesting was the use of two overhead projectors and one set of transparencies. So I followed his teaching method of explaining one transparency via one projector and having finished it, placed it on the second projector. I then explained the new transparency on the first projector. This way the students could check the previous transparency, if they needed to. This teaching method clearly made them very comfortable. It was also the course where student participation in the electronic chat room for the course was enthusiastic, that is, the one where I would step in if a discussion thread went astray. It was this course of eight lectures that at the end of it I received enthusiastic applause. I would note that I had actually modified some of my colleague's lecture materials notably by adding some transparencies with quantum mechanics equations. Maybe about a half to a third of the class had good enough maths to grasp it. But, like a public audience, it seemed that all the students appreciated my efforts to at least show them from where chemical orbital (electron density probability distributions) shapes arose from. This avoided the material otherwise appearing like a rabbit out of a hat!

A skill that you will need to plan for, although fortunately rare, is when technology fails. Universities provide a telephone line for instant technical support to arrive to help you. Naturally, this can take 5–10 minutes. In extreme cases, the rest of or indeed your whole lecture can be without your planned technology. Of course, you could abandon the lecture and try and reschedule it. However, I think the students appreciate your rising to the challenge. Indeed, if you follow the rules mentioned earlier, especially never overload too many slides or transparencies into your lecture plan, you can proceed with chalk and talk at the blackboard without electronic technology. Or if you feel comfortable and have some good questions, you can proceed to stimulate a discussion with the students.

No lecture course is without the support of teaching by tutorials, which you yourself will deliver, and quite often as well as workshops. These both reveal a great deal of the good, and the to be improved upon, elements of your lectures because you get informal feedback from your tutees. It is very interesting then to see the rise of the massive open online course, an Internet-based distance learning program designed for the participation of large numbers of geographically dispersed students. On my retirement, I have tried one of these as a student (introduction to philosophy) and gained a lot from it, although I did not complete it. There were many weblinks, most of which worked but some frustratingly did not. As I was not paying for it, I had no redress. Other distance learning situations I am familiar with involve industrial placement students (on a Chemistry with Industrial Experience degree course). These students being away on placement for a year would take two distance learning courses presented to them on CD. They were assigned a tutor who could respond by e-mail to their questions that would arise. The year's performance was assessed by a visiting tutor; I served for five years undertaking that role.

Overall, there is the simple maxim that you will sometimes have to explain to students that they have to make an effort themselves in order to learn.

REFERENCES

1. P. Day (2012) *On the Cucumber Tree: Scenes from the Life of an Itinerant Jobbing Scientist*. Grimsay Press, Glasgow.
2. J. R. Helliwell (2014) Book review of *On the Cucumber Tree: Scenes from the Life of an Itinerant Jobbing Scientist*. *Crystallography Reviews* 20(2): 157–159.
3. J. R. Helliwell (2009) Lecture demonstrations in a Public Lecture on X-ray Crystal Structure Analysis: From W L Bragg to the Present Day, *J Appl Cryst* 42, 365. Retrieved from http://www.iucr.org/education/teaching-resources/bragg-lecture-2001.

23 How to Supervise PhD Postgraduate Students

The nature of the supervision you provide to PhD students during their training considerably varies through their time with you, usually three to four years.

Obviously, the supervising of an experimental science PhD, which will still have or should have a strong theoretical basis, will be different from guiding pure theoretical/computational/simulation postgraduates. The latter will spend a lot of their time sitting in front of computer workstations, whereas experimentalists may be constructing apparatus or synthesising new chemical compounds or purifying biochemical complexes or preparing biological specimens. Health and safety, as I mentioned in the Preface, will also have very specific legal requirements, which I cannot cover here. I emphasise core aspects of PhD supervision in the following. Suffice to say though that experimental PhDs need to be trained before embarking on potentially hazardous tasks or using hazardous chemicals.

You will not be alone with your supervisory work, as there will also be an appointed adviser and, in some institutions, a moderator. The adviser would usually be from the same, or related, research field. The role of the moderator might be to draw on help and insights from a different area of research. The adviser and the moderator (and not the supervisor) would likely be those that interview the research student for assessing their end-of-year reports and make the key recommendation whether the student should proceed to the next year of research and study.

A research student's progression can in its core terms be described as follows:

Year 1: You will be providing a lot of specific training via your lectures on basic theory and principles of experimental practice in your field. The lectures in fact may well be more like tutorials if you take on just one new PhD student per year, which was largely my case. You will also be providing very detailed hands-on guidance with equipment and software for data analysis. You may rely to some extent on the more experienced PhD students and postdocs in your group, not least when you will be away travelling, but I would emphasise that it is your job to be in control of the details of the training of your students. A bright student may bring you new results in their first year. To pass to year 2, your PhD student will need to write up an end-of-year report including a science literature survey chapter or two, several chapters on research topics and most importantly a good research plan to take forward.

Year 2: To begin with, this will be a quite similar feeling to the previous months but with some added sparkle that should come from the coherence that the student should find from having written their end-of-year 1 report. This report will, as I have mentioned earlier, have been independently examined by one, perhaps two, of your academic colleagues. You will have likewise undertaken the same examination role for other PhD students in your department. Year 2, as it develops, will guide you as to the ability, the level of commitment and the energy of your current student. The research plan in the end-of-year 1 report will, or should, really develop. You may find, however, that your student needs careful guidance at an almost daily level; if so, you should provide this in your role as a science research educator. By year 2, you will also have had chances to send your student away on weeklong, typically, national or international training schools. This will entail you raising the monies for their registration, travel and subsistence. You may be lucky in your field to have choice, and some conferences can provide a nice mixture of skills workshops, as well as research-led microsymposia, which are also important for your student to attend. By the end of year 2, in your student's end-of-year report, there should now be quite detailed research results in the chapters, which will basically have the same headings as the end-of-year 1 report research plan. If this report requires fairly little by way of feedback, things are looking rosy for a good transit into year 3. Alternatively, a lot of interaction may be needed with the student, for example, via commented drafts; if so, you should still provide this in your role as a science research educator. The research plan for year 3 is a vital part of the end-of-year 2 report. The report should again be examined by one/two of your academic colleagues. You would again be doing the same role as examiner for other PhD students' end-of-year 2 reports in your department, and this is necessary for your broader perspectives.

Year 3: Expectations have changed somewhat over the decades. Funding agencies such as the Wellcome Trust and the UK's Biotechnology and Biological Sciences Research Council (BBSRC) have introduced funded PhDs spanning 4 and 3 1/2 years, respectively. Previously, all PhD studentships were funded for 3 years maximum. However, the lengthening of the PhD period is in response to the complexity of research and the techniques that must be mastered by the PhD student today. This is relevant to our discussion here as it shows the importance of patience on the part of your role as supervisor and educator as well as researcher. Individual academic research disciplines vary then according to whether you are engaged in biomedical sciences to those areas of science that still adhere to a three-year PhD. In both models, the university in the United Kingdom is formally monitored for how many students fail to submit their PhD thesis within a maximum of 4 years. It might well be the case, if longer than 3 years is needed, that your student is working without payment during their final year or final six months. Later, after their graduation with a PhD, you will get reference requests when you will be asked questions such as: what percentile (for example top 10%, top 25% etc.) had your student performed at during their PhD research and studies. Indeed it is in their final year of their research with you that you really learn about your supervisee's capacity to make their own mark in research as a career or not. Objective measures, like whether they presented a talk at a conference (and how many times they had to practise it in front of

FIGURE 23.1   As a supervisor, you will need to let your students take risks, even if it may leave you shouting 'Be careful' at times. (Courtesy of *The Times*.)

you and your research group), will be available to be evidence based in your writing of references. Also, there will hopefully be publications they have contributed to. But more subjectively, you know how much help the student had needed with the writing of their thesis chapters.

Overall, PhD students are not or should not be your research workforce. You will work together nicely with good students on research, and so you will almost not notice the difference between them being a research worker or a student, especially in their final year. Other students that you will supervise may, pretty much all the way through, remain your students. Supervision and training of students are clearly a vital role that you take on as an educator. Recall that you were young once and you are now part of that future of your research discipline because of the good supervision you received. I had a great PhD supervisor, Dr Margaret Adams, chemistry tutor at Somerville College Oxford University and researcher at the Laboratory of Molecular Biophysics, and I pay tribute to her as my DPhil supervisor here in these pages. She trained and guided me, yes absolutely, but she also gave me freedom* to develop my own scientific interests and skills and take risks (see, for example,

---

* One of my past PhD students, Dr Titus Boggon, now at Yale University, in reading my draft of this book, reminisced about this himself and wrote, 'This is absolutely what you did with your graduate students – I try to model myself after that, and I am grateful for the independence that you afforded me during my PhD. An exciting and formative experience!' It was a feel-good, proud moment to read that comment from Titus.

Figure 23.1). As well as this training, there is of course the formal expectation of the university that your student's thesis will be a contribution to new knowledge, and provided it includes a publication or two that will ease the assessment by the external and internal examiners of your student's thesis.

At the oral examination, the candidate will also need to demonstrate that they did indeed do the work that they have described in their thesis. You will of course have stressed to your student to never include anything that they cannot explain to their examiners.

A final point: How do you attract PhD students in the first place? Prospective students will contact you to make a visit to discuss possible projects with you. They will be assessing you as much as you will be assessing them. An excellent advert for your good PhD supervision and project quality is if your PhD students win poster prizes (Figure 23.2). Good results like that, with a picture of course, can be advertised on

FIGURE 23.2   Here I am presenting the IUCr poster prize at the British Crystallographic Association Biological Structures Group (BCA BSG) conference held at the Manchester Institute of Biotechnology, in December 2015, to Chris Earl, a fourth year PhD student at Birkbeck College, London University, and whose poster presented a new X-ray crystal structure of a protein's surface polypeptide loop inserting into the DNA double helix. (Courtesy of Dr Steve Prince, Dr Jordi Bella and Chris Earl.)

your laboratory notice board. Your prospective new PhD students can see that when they visit. In a delightful meet students situation, Sir Paul Nurse, past president of The Royal Society and Nobel Prize winner was asked, 'How do I choose a PhD supervisor?'. He provided excellent answers and I urge you to listen to him [1].

REFERENCE

1. Nobel Prize Inspiration Initiatives (2015) Advice from Sir Paul Nurse on choosing a PhD supervisor. Retrieved from http://www.nobelprizeii.org; https://youtube.com /watch?v=PeFR5z88eaA.

24 How to Be a Good Mentor

> Mentors are people who, through their action and work, help others to achieve their potential.
>
> **G. F. Shea [1]**

My relevant experience to mentoring, as distinct from other roles such as senior manager, line manager and supervisor, is that I had served as senior mentor for new academics in the School of Chemistry at the University of Manchester and a mentor on the University of Manchester Gold staff mentoring scheme for 10 years. In addition, in the scientific civil service, I was the director of a 240 person department. In academe, I have supervised 20 PhD students. I was also chair of the University of Manchester's School of Chemistry Athena Science Women's Academic Network (SWAN) [2] self-assessment team and gender equality champion, within which mentoring is a recognised useful approach to contribute to improving gender balance in science.

A second major category of one's mentoring is being on hand to offer advice to one's own past PhDs and postdoctoral scientists in their careers in science. The different stages of career are often referred to as *early career*, *mid-career* and *late career* respectively.

Arguably the most difficult career transition is to secure the first long-term contract (say spanning a period of at least four to five years), which is termed *tenure track*, and leads to the chance for the securing of tenure in academia. This career transition is most likely to come at the end of the second postdoctoral post, although sometimes it can be at the end of the first one. With the financial pressures on funding agencies and in turn on the universities and research institutes, the predominant posts available are not the longer term contracts but the two- or three-year postdoctoral ones. It is a major decision, not to be taken lightly, to take on a third postdoctoral post and can only be judged by the person involved as their circumstances can vary widely. To have the support of a mentor in this postdoctoral career period is important. I have therefore taken part in the University of Manchester workshops and mentor–mentee schemes to help postdoctoral scientists more widely than my own past PhDs and postdocs. I described details of the postdoctoral period in Chapter 1.

The Manchester Gold mentoring scheme's aims were to

> Provide a general framework for Mentoring of University staff in the Academic, Academic Related, Administrative and Research staff.

The scheme has been running since 2002. I participated as a mentor each year from 2003 to 2011 with one mentee per year. At my request, all but one of my mentees were based in the Faculty of Engineering and Physical Sciences or in the

Faculty of Medical and Human Sciences; one was from the School of Education. All mentor-with-mentee discussions are completely confidential. I can, however, offer some very general details:

1. Three out of nine were early career researchers (i.e. post-doctoral research assistants).
2. Four out of nine were mid-career staff who felt they had hit a career ceiling.
3. Two out of nine were staff who had recently arrived at the university.
4. One was an academic-related staff member (an experimental officer).
5. By faculty, four were from Life Sciences (including Pharmacy); three were from Engineering and Physical Sciences, one was from Arts and Humanities and one was from Medical and Human Sciences.
6. By gender, four were male and five were female.

The primary questions of each of the following three core groups were the following:

1. Early career researcher: 'How do I get tenure?' Or 'Which next steps should I take to proceed towards tenure?'
2. Mid-career researcher: 'How do I break through the career ceiling I am facing?'
3. New staff member: 'This University seems big and rather daunting, how does it really work?'

At this point, let us first define the terms *mentor*, *line manager* and *coach* (Table 24.1). Let us now expand the details of what a mentor provides to a mentee:

- Role model
- Critical friend
- Sounding board
- Facilitator
- Encourager

In my discussions with my mentees, I was keen to explore what precisely were their generic skills and so we would discuss in detail:

- SWOT analysis
- Time management
- Risks management

TABLE 24.1
Defining Terms

Role	Focus
Mentor	Career development
Line manager	Tasks/Results (Performance development review)
Coach (Trainer)	Knowledge/Skills

What is the role of mentor not to be about? It is not about the following:

1. It is not about being prescriptive.
2. It is not a purely passive or reactive role; it can be proactive. This latter aspect in my view overlaps with what is expected of a coach. Hence, my emphasis on encouraging my mentees to engage with understanding the general skills I listed earlier.

There are some important practicalities for your mentor-with-mentee discussions:

1. You should meet on neutral ground, not in the mentee's department or in your own department.
2. In the Manchester Gold scheme, we would meet every month for a total of six occasions.
3. Mentees volunteer and are thereby generally seeking a mentor at a time when they are faced with a significant career challenge.
4. The mentor-with-mentee discussion is strictly confidential.
5. There should be no discussion of personal problems; only career challenges should be discussed (mentoring is not a counselling service).

It is interesting that given the low proportion of women in academic science (~15%), the majority of my science mentees (five out of nine) were female. I later learnt, as gender equality champion for my School of Chemistry, of the importance of mentoring for women's academic progression to improve [2]. The reasons for this are also described in the book by Virginia Valian [3]. Basically, mentoring assists women with knowing what is needed for their promotion, to assist them in developing their self-confidence and, most simply, to encourage them to try for promotion (men will apparently try for promotion even if they think they have a low chance to succeed). On a very specific, gender-related note, I would also mention a shocking experience I had as a tutor. I confidently enquired of each member of an undergraduate students tutorial group of mine, a very bright and accomplished group, what their after graduation plans were, and specifically wondered if they would go on to do a chemistry PhD. One of the female students replied, 'There is no point carrying on with a PhD in chemistry as there are so few women chemistry academics'. Initially lost for words, I decided to work to train two female postdocs in mentoring. They had gladly volunteered from within our Athena SWAN self-assessment team when I mentioned that answer from this female tutee of mine.

Overall, was I successful as a mentor? A measure of this would be if the mid-career researchers did break through their glass ceilings. Some did, but not all. Did the early career researchers secure tenure? Some did, but not all. Did I, or the scheme, have any complaints? None to my knowledge. Did any of my mentees get invited to highlight their experiences at the end-of-year Manchester Gold get together? Yes, one did; I had a very positive feeling of pride, I must say! For myself, I also had a very positive feeling about every one of my mentor-with-mentee discussions through all those 10 years.

An interesting development these last few years concerns the pressures on PIs to hold back results to make the big, high-impact paper. Such an approach is well beyond the criterion for when to publish of significance and novelty and runs counter to the responsibility of the principal investigator (PI) as a supervisor in publishing results reasonably promptly to allow a better career development for the students and the early career researchers involved in a piece of research. There is an important ethical question here. Such issues and trends, in the context of mentoring early career and indeed later career scientists, will need to be addressed on a case-by-case basis, to resolve this ethical question, and I would expect them to grow in number as pressures on PIs grow to publish only high-impact papers.

Note: In the United States, supervisors of postgraduate students are referred to as 'mentors', although the role of PhD supervisor is not understood that way in the United Kingdom and who is also involved with setting of objectives, like a line manager. The UK's typical PhD student also has an advisor as well as their supervisor (as I described in Chapter 23), and the advisor has a supportive role akin to a mentor as I have defined it in this chapter.

Later, after a postgraduate student has got their PhD, you can serve as a mentor through their subsequent career, as I described above.

REFERENCES

1. G. F. Shea (1994) *Mentoring: Helping Employees Reach Their Full Potential*. American Management Association. Retrieved from http://www.amanet.org/.
2. The Athena SWAN Charter (2005) Recognising advancement of gender equality: Representation, progression and success for all. Retrieved from http://www.ecu.ac.uk /equality-charters/athena-swan/.
3. Virginia Valian (1999) *Why so Slow: The Advancement of Women*. MIT Press, Cambridge, MA.

Section V

Skills for Realizing Wider Impacts

25 How to Reach Out to Wider Audiences and Explain Your Research

In my recent book [1], I had a chapter entitled 'Explaining "what is crystal structure analysis?" for a general audience'. That chapter was of course quite specialised to my research speciality of crystallography; see Figure 25.1 for a spectacular general example! Here I will discuss some general aspects with details derived from the following:

1. My participation in university open days meeting prospective undergraduates, encouraging them to apply to my own university department and answering any questions from them or their parents
2. Lectures I gave to schoolchildren
3. Lectures I gave to the public, for example, at local community centres
4. Lectures I had given not only to wide interest groups of scientists from many fields but also to other people curious about science topics, for example, to the Manchester Literary and Philosophical Society with the University Physics Department (Schuster Laboratory) on the occasion of the university's celebrations of its 150th anniversary [2] and also presentation of a Friday evening discourse to the Royal Institution members in London
5. Synchrotron facility newsletter articles on prominent research results; these are newsletters that must reach out both to not only specialists and to the public especially but also including politicians, the pay masters of much of science
6. A plan I put forward to those responsible for education in UK prisons for me to present a lecture to prisoners as a contribution to their rehabilitation into society

There are common aspects to such outreach presentations:

1. The need to explain really quite complex topics
2. The need to avoid technical jargon wherever possible
3. Being a science subject, use as many demonstrations in one's lectures as possible
4. In the case of lectures, allow your audience to handle items you would exhibit either before, but usually after, one's lecture

FIGURE 25.1 Volta demonstrating his battery pile to Napoleon in 1801. I came upon this picture during my visit to the Volta Museum in Como, Italy; origin of the picture is unknown.

5. Give sufficient time for schoolchildren and members of the public to quiz you in detail directly; do not rush off

6. Not only prepare your lecture and exhibits carefully but also be in a relaxed state, for example, swimming earlier that day was a method I used to be on my best form possible as the lecture is as much about contact with you as the content of your presentation

7. Your exhibits need to have clear captions printed out in large text (your audience may be quite old and their eyesight may need to be assisted)

8. Your audience may ask you about very diverse scientific topics, which you could not necessarily know the answers to, so do not be afraid to say 'I honestly don't know!' In some cases though, you may be interested to find out and e-mail them your answer afterwards; this can be quite time consuming but helps your own wider education too. In Appendix 11, I offer detailed suggestions on how you could answer the very basic question 'What is the scientific method?', which can arise

A distinct way of reaching out to wider audiences is via the media such as TV, radio and newspapers. It is very likely that your institution has a media or a press section with expert staff to advise you. First of course is the question whether a research discovery is likely to be worthy of a press release and all the attention (phone calls, interviews with you and the like) that will flow from the press release announcement. Common sense tells you to avoid overhyping your results' implications. To make academic research accessible in nontechnical jargon will, however, test your skills of clear writing. You will in any case be assisted by your media officer. I give you a recent example of my own below [3]:

Example 25.1: Lobster Colour Change Mystery Solved

30 Apr 2015

For the first time scientists have come up with a precise explanation for why lobsters change colour from blue/black to red when cooked.

And the findings could have uses in the food industry and even in the delivery of some anti-cancer drugs.

When alive and living in the sea, lobsters are naturally a dark-blue/black colour. It is thought that natural selection led to this as it makes them harder to spot for predators. But put them in a pan of boiling water and they soon turn the familiar orange-red that is the colour that most people think for lobsters on their dinner plates.

Now scientists from the School of Chemistry at The University of Manchester with their international collaborators have come up with a precise explanation for why, after years of academic discussion. In a paper published in one of the journals of the Royal Society of Chemistry the team describe their findings.

The key is a chemical called astaxanthin, which has the orange-red colour of a cooked lobster, and how it interacts with a complex of proteins called crustacyanin which lobsters produce. The reaction of astaxanthin with the protein complex gives the creature its dark blue colour. But crucially, when the lobster is cooked, the protein is denatured and the astaxanthin is released and reverts to its orange-red state. This much was known from the X-ray crystal structure published by members of the same team in 2002.

The remaining scientific issue was the mechanism underlying the colour shift of free to bound astaxanthin. This is a much more complicated question! The clue is that astaxanthin can behave as an acid, a property that has emerged as important when it reacts with the lobster's crustacyanin proteins. It is this that the new publication has discovered to create the blue colour.

Professor John Helliwell of The University of Manchester, who led the team which carried out the research, said 'Over the last thirteen years since our X-ray crystal structure there have been competing groups studying this coloration mechanism, but hopefully now the issue is solved. It is a scientific curiosity, but it may also have important applications in the real world'.

'The coloration is quite a complex process to do with the 3 dimensional structure of the proteins in complex with the astaxanthins it binds, and the implications could be very useful'.

For example astaxanthin is an antioxidant, so it has many health properties. But because it is insoluble in water the problem is how to deliver it to a target. But our findings suggest that mixing it with crustacyanin could do that and allow the astaxanthin to get to a target such as via the stomach'.

It could also be used as a food dye, for example to help create blue coloured ice cream. Or it could be used in food stuffs to help people know when food has been cooked properly; a dot on the food that changes colour when it reaches a certain temperature could be used'.

'Most fundamental of all is arousing the curiosity of children and the public in basic science and our marine environment. In the era of climate change it is important for all to think about the delicate nature of life and the sustainability of life on the planet. How and why has lobster evolved this elaborate and delicate coloration mechanism? It is a beautiful and yet intriguing phenomenon'.

NOTES FOR EDITORS

The paper 'On the Origin and Variation of Colors in Lobster Carapace' has been published in *Physical Chemistry Chemical Physics*, a journal of The Royal Society of Chemistry DOI: 10.1039/c4cp06124a
http://pubs.rsc.org/en/Content/ArticleLanding/2015/CP/c4cp06124a#!divAbstract

MEDIA CONTACTS

Media Relations Office
University of Manchester

Extensive descriptions of explaining your research to various audiences and the public are described in Dennis Meredith's book [4]. Making yourself clear is the mission of Cornelia Dean's book [5], and in particular, Chapter 6 explains how to prepare for when a reporter calls! Prepare your top message (or two, and at the very most three, messages) carefully being her motto, and do not expect to be able to proofread the reporter's subsequent newspaper article draft!

REFERENCES

1. J. R. Helliwell (2015) *Perspectives in Crystallography*. CRC Press, Taylor & Francis Group, Boca Raton, FL.
2. J. R. Helliwell (2009) Lecture demonstrations in a public lecture on x-ray crystal structure analysis: From W. L. Bragg to the present day. *Journal of Applied Crystallography* 42: 365. Retrieved from http://www.iucr.org/education/teaching-resources/bragg-lecture -2001.
3. A. Haworth (2015) Lobster colour change mystery solved. University of Manchester, Manchester.
4. D. Meredith (2010) *Explaining Research: How to Reach Audiences to Advance Your Work*. Oxford University Press, Oxford.
5. C. Dean (2009) *Am I Making Myself Clear? A Scientist's Guide to Talking to the Public*. Harvard University Press, Harvard.

26 How to Handle Your Inventions, Patents and Services to Industry

A patent, or invention, is any assemblage of technologies or ideas that you can put together that nobody put together that way before. That's how the patent office defines it. That's an invention.

Dean L. Kamen
*An American entrepreneur and inventor (born 1951) best known
for inventing the electric self-balancing human transporter
with a computer-controlled gyroscopic stabilization and control system*

Whilst Marconi made a transatlantic transmission on 12 December 1901 Nikola Tesla's system was responsible for transmission before Marconi and was successfully used in Portugal to communicate over the Atlantic. As early as 1892, Tesla created a basic design for radio. On 8 November 1898 he patented a radio controlled robot-boat. Tesla used this boat which was controlled by radio waves in the Electrical Exhibition in 1898, Madison Square Garden.

Tesla Memorial Society of New York [1]

There is no patent. Could you patent the sun?

Joseph Salk (1914–1995)
Discovered the vaccine for polio in 1955

Who owns science?

The Manchester Manifesto [2]

Patents can lead to disputes, as the famous story about Nikola Tesla above is an example. Patents need not be pursued either as Joseph Salk exemplified. Indeed, 'who owns science?' [2]. The Manchester Manifesto [2] assists us to develop our understanding of our core values as scientists, while respecting our rights as researchers and inventors, and gets to the multifaceted missions of the modern university:

ISSUES/PROBLEMS IN THE CURRENT MANAGEMENT OF INNOVATION

The interests and contributions of inventors and authors deserve to be recognised fairly. However, the current dominant model of innovation and commercialisation of science poses a number of problems. It has potential to encourage innovation and stimulate research and development, but also to frustrate innovation and stifle research and development; and can hinder science from operating in a way consistent with the public good. [2]

As scientists, we have to get used to the fact that we are more and more encouraged to step outside our ivory towers of quiet laboratory work and, in addition, define the possibilities for the intellectual property (IP) arising from our science discoveries. We must also lay out our plans for the knowledge transfer to industry of the research that we propose to undertake in our research grant proposals. Of course, industry was the traditional venue for such applied research work, while university scientists got on with predominantly curiosity-driven basic research. However, excellent basic research can take place in industry too. For example, the discovery of the active chemical agent of the female contraceptive pill was the result of the industrial laboratory research work of Dr Carl Djerassi, and the patent was granted to the company that he worked for at that time, Syntex Corporation, in Mexico. Carl Djerassi received pretty much every award in science except the Nobel Prize for that basic, then applied, research discovery in organic chemistry. E. P. Abraham as a research chemist at Oxford University discovered the antibiotic cephalosporin, similar to penicillin, and whose eventual sales income led to a fund, a registered charity [3] to

> Support education and research in the medical, biological and chemical sciences carried out in the University of Oxford, The Royal Society of London and King Edward VI School in Southampton to help the public.

What skills do you as a scientist need to have to take care of your IP? Do the research grant conditions on IP differ from funding agency to funding agency or from employer to employer?

The role of a university, with a diverse mission including basic research and IP protection, is perhaps now better organised with respect to IP matters than in the past. As a researcher, I have found excellent support within the University of Manchester when I had patentable research results or ideas. Overall, my efforts in this area involved only one patent during my research career, a one-year patent. This was a 'three dimensional detector for polychromatic X-ray Laue diffraction measurements especially suited to congested patterns of diffraction spots that would otherwise overlap too much in a traditional two dimensional detector'. The limited market of interested users of such a device meant that it was not worthwhile to pursue the great costs of a full twenty years duration, worldwide coverage patent. Nevertheless, it taught me a lot. The university's IP Support Office of the time were very helpful. Today, the equivalent section is called The University of Manchester's Agent for Intellectual Property commercialisation and is a division of UMI and is wholly owned by The University of Manchester, founded to commercialise relevant university research results. What is UMI? It is the university's innovation company, standing for *inspire, invent, innovate*.

Besides my new three-dimensional (3D) polychromatic X-ray diffraction detector, I had the occasion to consult UMIP on two other possible research areas with IP potential. I think the details are interesting to share with you as an illustration of two things: Firstly, in case study 1, respecting your research student's IP; secondly, recognising your moments of imagination, but which might come to nothing in the cold light of day of your IP support officer.

Case study 1: With a PhD research student, we were working on determining the crystal structure of a protein whose inhibition could have a useful role in slowing down tumour growths. The crystal structure determination was held up due to a type of crystal imperfection that can plague our studies sometimes. My research student decided to create a 3D model of the protein structure, a hazardous business as the closest known 3D structure only had a 20% amino acid sequence identity. Secondly, he modelled some existing known inhibitor compounds in the computer to seek improved compounds. These would perhaps bind more strongly. I had insufficient expertise to be of much help to him on such modelling, but the results certainly looked interesting enough to encourage him to discuss them with UMIP. I asked him if he would like me to sit in on the discussion, but it was up to him as it was his IP. He said that would be good. UMIP arranged a meeting with a specialist patent lawyer who had detailed experience of the pharmaceutical industry. The advice given to my student was that he would need to synthesise the compounds he had designed and assay them for activity against the protein. My student then arranged to meet a synthetic chemistry group at the nearby Cancer Research Institute. Again, there was insufficient evidence to proceed. We had already tried and failed to obtain research grant support within a broader research proposal. It seemed that we had done what we could to 'check out the IP aspects'. Nevertheless, I then encouraged my student to write up his study for publication. This was submitted under his name alone, obviously; I was thanked for discussions, which was appropriate. The editor rejected the article because my name was not on it as a coauthor. This was ethically quite ridiculous, and my student wrote to the editor, at my suggestion, as follows:

I am in safe receipt of the decision and reports on my submission 'xxxxxxxx'
    There is one point that arrested my attention, and that of my supervisor, Prof John R Helliwell:
    'It is also submitted by a student without the advisor listed as a coauthor. This has me rather concerned.'
    In response Prof Helliwell wishes you to know that:
    'The motivation for my student being sole author was simple. He had initiated that structure prediction and ligand discovery study! I am surprised that that option did not seem to occur to the Editor!'
Sincerely,

The detailed referees' reports in any case were much the same as the IP advice we got, 'The author needs to prove that his compounds really do inhibit the protein's activity'. The case study does illustrate though the importance of the supervisor respecting the sole rights of the student when such a situation arises, and which it can and did!

Case study 2: This involved one of those bursts of imagination moments. I wondered what if one could miniaturise a synchrotron to the nanoscale? A nanosynchrotron! This could be very versatile not least for academic or industry research staff to have their own personal synchrotron for their analysis needs; imagine having a nanosynchrotron pen in the top pocket of your lab coat, like a laser pen! This could be ready to be used whenever you wanted to undertake an analysis on your sample! My nanosynchrotron concept was to have a cyclic arrangement of manganese atoms in a chemical cluster, inject electrons into it and in going around the circular layout of manganese atoms, the electrons would radiate light, as in a conventional synchrotron. I chose manganese because I knew that manganese is very versatile with its (most common) chemical oxidation states: II, III, IV, VI and VII being viable. Thus, a chemical situation similar to the oxygen-evolving complex cycle of Mn II and Mn III in photosynthesis might be exploited. Let us not worry for the moment, I thought, about maybe only getting the electron(s) to go around the manganese cluster once or that it might only be one electron thus limiting its light emission intensity. So I carefully noted my ideas and nanosynchrotron designs in my lab notebook, signing the pages, and had a colleague countersigning them too. When I felt ready with my signed and dated notes, I contacted my UMIP support officer. The meeting was arranged and a preliminary patent search was conducted by her beforehand. This revealed that graphene had been reported in a research article to be the fastest conductor of electrons yet recorded. It was first of all interesting in that the paper was from my colleagues across the street in the Physics Department! Also secondly, it made me realise that while I might get electrons to go round in a circle in a manganese cluster, the speed at which they might travel would be too low to be of interest to realise the synchrotron-like emission of radiation (the electrons need to be accelerated very close to the speed of light limit). The graphene research paper I mentioned earlier is an interesting example to compare with. The case study illustrates several points: respect your imagination, carefully document your ideas and designs and finally do contact your IP support officer. Of course, if it does not work out, you should buy your IP support officer a bottle of something for Christmas!

So today, the role of the university scientist, and the skills we need, can, and perhaps must, include the roles of inventor and innovator to be added to our repertoire of educator, teacher and researcher. These are diverse times for us as scientists. As I say, not that this did not happen before, but that there is a growing expectation from government, industry, the universities and indeed the public that all scientists need to be aware of and trained in these activities. A general analysis of the interrelation of the modern university, industry and government is described in the *Triple Helix* by Henry Etzkowitz [4]. What are the golden rules that I have discovered about this activity so far in guiding my work? Most importantly, if you have an idea that your research may have IP potential, do not reveal your research results until you have consulted your university's IP department; do not present a seminar or a lecture or submit an article before then! Let us come back to ethics. Of course, if I were looking at a vaccine for polio like Joseph Salk, how could one hold that research result back behind a patent?! It would surely contradict one's own ethics and principles to not openly reveal your results. Submission for publication, and awaiting peer review and acceptance, would also involve delay and in such a case, deaths of people not given the vaccine. But what if the results had a flaw that peer review might pick up? That would justify the delay in announcing your results via a press release. A colleague of mine had important crystal structure results of a protein whose inhibition could stop bird flu. Time was of the essence. The IP could be protected without too much delay but would require a patent lawyer's professional assistance on a Sunday. A patent lawyer in Israel, who worked on a Sunday, was duly consulted. The research article was submitted to a journal on a Monday. There are ways then in our diverse and fast communication world to cover all the bases that we, as scientists–employees, have to follow and yet proceed promptly!

In terms of analytical services to industry, I was involved in detail with one of these, namely at the highly versatile source of radiation, the UK's synchrotron radiation source (SRS). As a beam line scientist in the early 1980s, I costed the beam time access for the pharmaceutical industry to take advantage of the spectacular properties of our X-ray beams for protein crystallography. This became a highlight in the final overview analysis, after the closure of the UK's SRS in 2008 [5]. An increasing number of large-scale synchrotron radiation (SR) facilities exist worldwide. The UK's SRS [5] provided state-of-the-art analytical techniques from infrared to hard X-ray wavelengths. The unique characteristics of SR are ideal for analytical problems that require high spatial or temporal resolution or problems that are simply intractable using conventional instruments. The SRS and other large facilities have been traditionally used by universities and higher education institutions for pure research and development. In recognition of the needs of commercial customers, Daresbury Laboratory established the Daresbury Analytical Research and Technology Service in the mid-1990s as a 'materials characterisation service to a wide variety of industries' such as involving pharmaceuticals, biotechnology and healthcare, chemical, materials, petrochemical, automobiles and aerospace, instruments, electronics and photonics. The successor to the SRS, the Diamond Light Source, with hugely enhanced technical capabilities over the SRS, continues this very active analytical services to industry tradition; see http://www.diamond.ac.uk/Industry/.

An example of the enthusiastic support for science from the government is the then UK Prime Minister Gordon Brown's 2009 Romanes Lecture at Oxford University, where he cogently argued that investment in science and the next generation of scientists is key to the UK's future competitiveness [6].

REFERENCES

1. Tesla Memorial Society of New York (n.d.) Nikola Tesla is the father of radio, not Marconi. Accessed 7 August 2016. Retrieved from http://www.teslasociety.com/tesla _against_marconi.htm.
2. J. Harris and co-signatories University of Manchester: Institute for Science Ethics and Innovation with the Brooks World Poverty Institute (2009) *The Manchester Manifesto.* Accessed 7 August 2016. Retrieved from http://www.isei.manchester.ac.uk/The ManchesterManifesto.pdf.
3. E. P. Abraham (1970) *The EPA Cephalosporin Fund.* Accessed 7 August 2016. http:// opencharities.org/charities/309698.
4. H. Etzkowitz (2008) *The Triple Helix: University-Industry-Government Innovation in Action.* Routledge, New York.
5. UK Science and Technology Facilities Council (STFC) (2010) *New Light on Science The Social & Economic Impact of the Daresbury Synchrotron Radiation Source, (1981–2008).* Accessed 7 August 2016. Retrieved from http://www.stfc.ac.uk.
6. G. Brown (2009) Science and our economic future. The Romanes Lecture, Oxford University, Oxford. Accessed January 2016. Retrieved from http://podcasts.ox.ac.uk /gordon-brown-science-and-our-economic-future.

27 How to Help Improve Gender Equality

Is it wholly Utopian to suggest that women scientists and science teachers who require leave of absence for family reasons should be given at least one year's paid leave for each child as a matter of course, or that Universities, local authorities and other employers might expect to provide well-equipped nurseries and kindergartens just as they provide refectories and common rooms? ... (They might even sometimes be useful to young fathers especially to the students whose wives are working while they study!).... The kind of woman, like Professor Dorothy Hodgkin, who has managed to win a Nobel Prize for Chemistry and to bring up a family of healthy, intelligent and socially-active children, is a tremendous inspiration to other young women.... Encouragement (also) means a tremendous amount to a young woman.

Kathleen Lonsdale
President of the International Union of Crystallography 1966
Extract from her Royal Institution Friday Evening Discourse
transcript January 23, 1970 entitled 'Women in Science' (p. 317)

To be in a university science department comprised only of men, which I saw in three different universities, two of physics and one of chemistry between 1979 and 1989, is so statistically unlikely as to warrant the questions why and how. And yet when I chose my DPhil project at Oxford University in 1974, my chosen project's owner was a woman scientist, Dr Margaret Adams. I had not thought about the gender of the supervisor; I picked the project. I presume therefore that *unconscious gender bias* did not apply to me. There may be unconscious bias that would explain the lack of women in those three science departments that I mentioned earlier, only that I firmly believe I have not been infected with it. So have my colleagues been infected? Or is there one or more other forces at work that would explain this gender inequality in the science workplace? Is science a unique owner of the problem?

In 2007, I was asked by the School of Chemistry head of department at the University of Manchester to be the school's gender equality champion. I was not quite sure why I had been asked. Maybe, as one colleague observed, I was simply 'a very approachable person to students and staff alike'. Or it was because crystallography as a field had a good reputation for a better gender equality. Both these theories are plausible I think.

First, let us report the findings of fact on the gender balance in academic science departments. Figure 27.1 shows the balance at different stages of career progression. Figure 27.2 shows the gender balance across the countries of the globe. These figures show the overall picture in science. There are significant variations by science subject. As our department gender equality champion, I was able to study the best

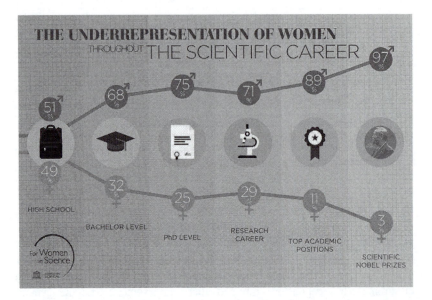

FIGURE 27.1 The gender balance at different stages of science career progression; these data reveal to all of us that a multiplicity of different types of actions is needed to stop the percentage drop of female staff at the various stages of an academic career. (Courtesy of the L'Oréal Foundation for Women in Science.)

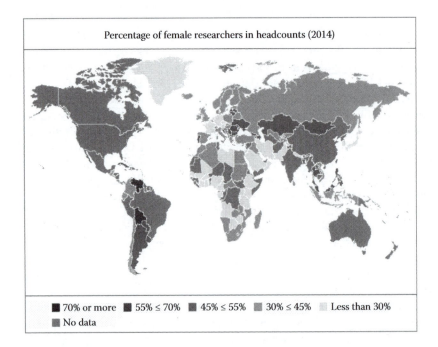

FIGURE 27.2 The science gender balance across the countries of the globe. (From UNESCO Institute for Statistics, http://www.uis.unesco.org [6]. With permission.)

practice of various other academic departments in the United Kingdom because the documentation of such departments, registered as bronze, silver or gold standard with the UK's Athena Science Women's Academic Network (SWAN) scheme, had to be open access. So while chemistry and physics typically had overall maybe 10% female academic staff; a biology department would have 40%. With that said, this stemmed from a typical undergraduate biology class with, approximately, 60% female students. Nursing, I learnt, would have 85% female academic staff; the gender balance issue here was whether a better male percentage was appropriate.

Both Figures 27.1 and 27.2 show the hard facts that there is a gender balance problem in science careers. To address this major and very important problem, organisations and existing staff need to address what skills they are missing or are poor at in the recruitment and the retention of women in science. To this end, the UK's Department of Business, Innovation and Science established the SWAN Athena Project [1], whose aims are

… the advancement and promotion of the careers of women in science, engineering and technology in higher education and research, and to achieve a significant increase in the number of women recruited to top posts.

The inaugural SWAN was founded at Imperial College by Professor Julia Higgins, FRS and her colleagues.

The European Research Council (ERC) has also taken affirmative action and declared that it will

Raise awareness about its gender policy among potential applicants and improve the gender balance among researchers submitting proposals to the ERC in all research fields.

Identify and challenge any potential gender bias in its evaluation procedure and monitor possible changes in gender structures (careers and academic posts) following (the award) of its grants.

Achieve in the medium term gender balance among its peer reviewers and other relevant decision making bodies, with a minimum participation of 40% of the under-represented gender.

The problem of the underrepresentation of females is also seen as a real problem in business at the top levels; a high-profile report is that from Lord Davies entitled 'Women in the Boardroom' and whose declared aim was to 'reach 25% female representation in the Boardroom within 3 years or face quotas' [2].

For science departments, to return to our focus, the declared options at the University of Manchester to address this state of inequality are

The University can train employees for work in which they are under-represented.
The University can encourage under-represented groups to apply for posts.

Within this overall framework, I set about my role as the School of Chemistry's gender equality champion and, with my fellow committee members of our Athena SWAN self-assessment team, set about raising awareness within our staff that

The School is very aware of the important female talent pool in chemistry, notably in the excellent success of our female chemistry graduates, who comprised around 45% of our graduates, and which needs to translate into a yet further improved female to male academic staff percentage split if the School is to compete with other leading chemistry Departments around the World in the future.

We need to be fairer to the female graduates and staff group; this is an extremely serious issue of inequality.

As practical steps, we led the School of Chemistry to engage

- Yet more closely with the Faculty of Engineering and Physical Sciences Human Resources Team to benefit from Equality and Diversity Training
- With other science departments such as Physics and Chemical Engineering and Analytical Sciences (CEAS) in aiming to share best practice towards fully reaching the Athena Swan, UK national level principles of best gender equality practice
- With the Faculty of Engineering and Physical Science Women in Science, Engineering and Technology (WiSET) network, which was for female students, research and academic staff in the Faculty
- With Athena SWAN nationally to achieve best practice, aiming ultimately for an Athena Swan 'Gold Award' standard of gender equality at all levels of our academic staff

Overall, we have gained successful improvements (see Figure 27.3).

We also found that the Royal Society of Chemistry (RSC) had developed a gender equality guidance booklet and self-assessment tool [3]. The self-assessment team found out that about a quarter of the points listed in the self-assessment tool referred to checking of opinions of academic staff, and which therefore required a staff questionnaire

FIGURE 27.3 On behalf of the School of Chemistry, University of Manchester, we accept our Silver Athena SWAN Award from Professor Julia Higgins, FRS, patron of Athena SWAN. From right to left: Dr Anna Valota; Professor Julia Higgins, FRS; Dr Cinzia Casiraghi; and myself. The event was held at Imperial College London. Dr Valota served as chair of our School of Chemistry Post Docs Forum and member of our school board.

to be designed, which we did ourselves since at that time there was no institutional or national-designed, standard questionnaire. Also, there were questions we could starkly see that we had poor performance on: the poor timing of our seminars, the very low percentage of female speakers, the very low fraction of females on our main committees, and the very low number of senior female academic role models for our younger female chemists. So we set about trying to secure major improvements. In the RSC's assessment tool, we also found a very difficult group of questions relating to achieving a gender-balanced short listing for academic vacancies. This was due to the worldwide nature of the gender imbalance in science (Figure 27.2), where there was a genuine low proportion of female candidates, and so search committees as well as our Human Resources Department should, in looking out for the best applicants for a vacancy, would, most likely, need to extend their searches onto a European/global level.

Overall, although much has been set in motion, progress is anticipated to be slow; a metric is an estimate for each academic subject of which year gender parity will be reached and presents very sombre reading. Indeed, *Why so Slow* is the title of the extensive analysis in the book by Virginia Valian [4], and I cited her book in Chapter 24 on mentoring. Virginia Valian's analyses span not only academia but also into other professions such as medicine and law. She also provides incisive examples of interviews of individuals and provides an extensive set of references. These include studies of how many publications are published, and how well cited these are, for cohorts such as women without children and women with children, compared with men. Women, without or with children, tend to have similar publishing habits (publishing fewer articles on average than their male counterparts) and achieve higher average citations than men. This is all a very large area of contemporary study. Suffice to say that the guidance in practical terms from schemes like Athena SWAN and professional bodies like the RSC are vital to individual departments to progress to a better and fairer future.

L'Oreal-UNESCO also offers excellent practical advice with its 'Commitments for Women in Science: A Manifesto for Change' [5]:

1. Encourage girls to explore scientific career paths.
2. Break down the barriers that prevent women scientists from pursuing long-term careers in research.
3. Prioritise women's access to senior positions and leadership positions in the sciences.
4. Celebrate with the public the contribution that women scientists make to scientific progress and to society.
5. Ensure gender equality through participation and leadership in symposia and scientific commissions such as conferences, committees and board meetings.
6. Promote mentoring and networking for young scientists to enable them to plan and develop careers that meet their expectations.

REFERENCES

1. UK's Athena SWAN Project (2005). Accessed 7 August 2016. Retrieved from http://www.ecu.ac.uk/equality-charters/athena-swan/.
2. Lord Mervyn Davies (2011). Accessed 7 August 2016. Retrieved from http://www.bis.gov.uk/news/topstories/2011/Feb/women-on-boards.

3. The Royal Society of Chemistry, *Planning for Success: Good Practice in University Science Departments*, Burlington. House, Piccadilly, London.
4. V. Valian (1999) *Why so Slow: The Advancement of Women*. MIT Press, Cambridge, MA.
5. L'Oreal-UNESCO (1998) *Commitments for Women in Science: A Manifesto for Change*. Accessed 7 August 2016. Retrieved from http://www.fwis.fr/?locale=en, launched in 2015.
6. UNESCO Institute for Statistics (2007) Report on *Science, Technology and Gender: An International Report* by Walter Erdelen. Accessed 7 August 2016. Retrieved from http://www.uis.unesco.org.

Section VI

Leadership Posts

28 How Do You Know If You Really Want to Be Head of a Department

Let us start with an example, from the then head of the United States' President John F. Kennedy:

> I believe that this nation should commit itself to achieving the goal, before this decade is out, of landing a man on the moon and returning him safely to the earth. No single space project in this period will be more impressive to mankind, or more important for the long-range exploration of space; and none will be so difficult or expensive to accomplish. We propose to accelerate the development of the appropriate lunar space craft. We propose to develop alternate liquid and solid fuel boosters, much larger than any now being developed, until certain which is superior. We propose additional funds for other engine development and for unmanned explorations – explorations which are particularly important for one purpose which this nation will never overlook: the survival of the man who first makes this daring flight. But in a very real sense, it will not be one man going to the moon – if we make this judgment affirmatively, it will be an entire nation. For all of us must work to put him there. (Excerpt from an address by former US President John F. Kennedy before a joint session of Congress, on 25 May 1961.)

And the goal was realised:

> That's one small step for man, one giant leap for mankind. (Words said when Neil Armstrong, astronaut on the USA Apollo 11, first stepped onto the moon [20 July 1969].)

The quotations from John F. Kennedy and Neil Armstrong are extreme examples of leadership and realisation of an overall goal. But they illustrate the importance of a vision espoused by a leader with the power and resources to effect a vision, and with the necessary expertise in the teams needed to carry out that vision. In this and the next two chapters, I explain three different kinds of situations where you can effect leadership. Two chapters deal with leadership at senior management level, and the third deals with leadership at a more junior level, but is leadership. Figure 28.1 is a sketch of the personal qualities of a good leader.

As your career develops, you may have an eye on becoming a head of a department. When I was 49 years old, I had this gleam in the eye and applied for the post of director of synchrotron radiation science at the UK's SRS at Daresbury Laboratory, near Manchester, and within which role I was head of the synchrotron department responsible for UK's SRS and its science. I also provided direct support, with my department staff, to the new UK Diamond Light Source Project at the Rutherford Appleton Laboratory near Oxford.

FIGURE 28.1   Qualities of a good leader.

The qualifications for a head of an academic science department that you will be asked to document will likely include the following:

Qualifications

1. A PhD in a relevant scientific/technical field
2. A minimum of 15+ years experience
3. Comprehensive knowledge of, and a high scientific reputation in, the application areas of the department
4. Experience in managing and maintaining effective interpersonal and external stakeholder relationships
5. Experience in managing safe, efficient operations in the department
6. Experience developing, managing and motivating a high-performing and diverse population of scientific, technical and professional support teams
7. Experience of staff equality and diversity
8. Experience in managing large projects, in scale and in budgets

The skills that you may well be asked to have will, for example, be as follows:

1. Demonstrated ability, drive and organisational skills to contribute to research leadership and to provide leadership to colleagues in the delivery of high-quality research and teaching
2. Ability to work independently and as part of a team
3. Ability to provide academic, strategic and operational leadership
4. Excellent oral and written communications skills
5. Excellent interpersonal and presentational skills and ability to enthuse students
6. Proven ability to deliver undergraduate teaching at all levels
7. Ability to develop taught courses and to teach effectively at undergraduate and taught postgraduate level

8. Ability to lead curriculum development in well-chosen topics with wide international appeal
9. People, time, project and budget management skills as appropriate
10. Ability to organise and encourage others to work effectively and independently with little supervision
11. Proven research creativity with an ability for independent thought required to generate original projects leading to grant awards
12. Ability to accept collegiate responsibilities and to act accordingly
13. High degree of self-motivation and vision
14. Ability to provide wide academic managerial or administrative leadership

So if you think you tick all the boxes for such a job advert, you might first ask, 'What would be the risks that I would become risk holder of?' It is the risks folder that you would assume ownership of that would potentially keep you awake at nights, and it would be best that you would find out in advance of signing on the dotted line.

My role as a head of a department, which I mentioned at the beginning of this chapter, lasted a year. It ended when, overwhelmed with anger about the budget discussion for my department, I gave the chief executive my resignation. I later tried to withdraw my resignation but he said, 'Oh no, you have resigned, you are far too interested in science'.

29 How to Lead Your Learned Society If You Are Elected as Its President

This is a major commission that may happen to you if your colleagues decide to put your name forward and you succeed in the election. This will only have happened if you are well known to your colleagues and you will know well the workings of your learned society. You will have a committee of eager, well-qualified, elected colleagues on the society's executive committee. There will be a suite of regular tasks ahead of the committee such as the next few conferences of the society, which is likely to be delegated to a subcommittee or to the local organisers selected in some way. So as leader, its president, what do you focus on? You will of course respond to queries that will come to you from your colleagues. You may be concerned over this or that activity of your society and you will ask pertinent questions of the elected officers. Overall though, what do you personally do?

In my case, I was elected president of the European Crystallographic Association (ECA) in 2007 and served for the usual three years and then a further three years as its past president, making six years in total on its executive committee. I had previously been a founding chair of a special interest group, a SIG, and served as chair for five years. The SIG I had responsibility for was Instruments and Experimental Techniques. In that role, I represented the SIG to the ECA council, the overarching body of some 30 national councillors, and three individual members' representatives, adding a further three councillors. The executive committee would prepare matters or issues for voting at the ECA council. In between the annual council meetings, councillors could be consulted by e-mail and undertake an e-mail vote as may be required. Overall, the ECA geographic territory for its members and its conferences covered all of Europe and Africa (until such time as Africa would be ready to form its own organisation for crystallography). The statutes of the association covered the following:

- To foster and encourage the growth and the application of crystallography
- To uphold and advance the standards, the competence and the conduct of our science
- To serve the public interest by acting in an advisory, consultative or representative capacity relating to crystallography
- To advance the aims and objectives of our members so far as they relate to the advancement of crystallography or the practice of crystallography

I decided I could do one new thing for the first executive committee meeting that I would be chairing, which was held in Budapest in February 2008, and that was to prepare a talk on a future strategy for the ECA. The contents of my talk were the following:

- Overview
- Our SWOT analysis
- The association as an international organisation
- Developing the association; financial objectives and membership
- Environmental requirements for a modern organisation
- Ensuring the next generation; the three Es, that is, education, education, education
- What the association should not do
- Measuring the association's impact; agreeing the key benchmarks
- Establishing and diligently managing the association's risks

In the overview, I covered how we would continue our work in concert with the European and African national societies and add value; that is, we, the ECA, do not plan to replace the national associations. We would organise the professional development (CPD) via our conferences (the European Crystallographic Meetings [ECMs]) and training workshops just before ECMs, foster the study and detailed understanding of crystallography via schools (a new venture), and seek to increase our influence on key decision making bodies (e.g. via the Innovation for Science in Europe organisation).

In a SWOT analysis, I felt that our association's strengths were that the ECA had a recognised presence (brand) notably via its ECMs; it funded conference bursary applications (as well as ECMs to a wide variety of conferences on crystallography in Europe and Africa); the ECA had several income streams. Our weaknesses I felt were that we had no permanent administrative staff support, and so continuity of operations was made more difficult, which were handled by the elected volunteers who would of course change over regularly; there were insufficient financial reserves. In terms of opportunities, I imagined that we could enhance the contacts with our global organisation, the IUCr, by inviting an IUCr executive committee link person to our ECA executive committee meetings. In terms of threats, crystallography as an academic discipline was being phased out in university undergraduate courses, and so the ECA needed to create and/or support education initiatives to deal with this threat.

In terms of ways we might develop the ECA's financial strengths from its then current level of assets of ~50,000 Euros, I outlined that we could make it a plan to grow our assets at some percentage per annum; we could adopt an overall target for an investment account capable, via the annual interest, to pay for the associations' prize awards and pay for the various bursary monies we awarded for different conferences and indeed plan to steadily increase these bursary funds; we could aim for a particular overall target value of say 250,000 Euros of assets (which at 4% per annum would yield an annual income of 10,000 Euros), and then we could readily accept more bursary award obligations!

We would aim to develop the number of our individual members and our corporate affiliate members.

In terms of recognising our environmental obligations for a modern organisation, we should encourage all efforts to protect the environment and to tackle climate change; taking many small steps could make a decent-size leap, for example, minimise the number of printed abstract books at ECMs and look into organising e-conferences (known these days as webinars).

Since the challenges of education were identified in the SWOT analyses as a threat, education initiatives should be given a high priority. So, for the postgraduate level, the ECA should strongly support the national initiatives and seek to initiate a European Masters in crystallography qualification. At the undergraduate level, the ECA should, via the IUCr Teaching Commission, propose an electronic archive of teaching PowerPoint Slides. For schools, ECA should find a way to support essay competitions.

It is always worth discussing what an association should not do. I put forward that we continue to not replace the national societies in Europe and Africa but help them.

In order to measure our associations' impact, we should monitor the following. Each year, did our assets grow adequately towards our overall target of 250,000 Euros? And each year, did our number of individual and corporate members grow adequately towards our targets? At five yearly intervals, supplement the annual evaluations of strategy by reassessing the long-term strategy.

The association naturally faced risks, and so the executive committee should agree to a risks folder, which should be reviewed at least annually, if necessary introducing new risks and/or deleting expired risks. The committee should reaffirm the risk holders annually.

I acknowledged to my executive committee colleagues the Royal Society of Chemistry's document 'RSC Strategy 2006–2010'. (Unfortunately, this is no longer accessible on the Internet.)

Overall, I believe that I made a fresh contribution to the association to plan better and more strategically for its future. Later, I became aware of two other associations reviewing their strategy. One had used a firm of outside commercially oriented consultants. Another association had consulted a senior colleague from another academic field but sufficiently closely allied to it. With the former case, with the commercial consultants, I felt the association concerned got a relatively generic analysis out of the process but it was OK. The latter case I felt was not good for the association concerned as the main recommendation was for the association to seek an alliance with a much larger scientific body but which had some years earlier launched an activity within its members of something akin to a SIG, similar to the smaller association. Subsequently, following the advice of the academic consultant, the smaller association suggested to the larger association to jointly organise a major conference. The much larger association, with a large administrative staff, would take the lead on the practicalities. So, to me, there was nothing good about that as a strategy as the smaller association had basically declared itself no longer needed!

In summary, I heartily commend to you to take your chance and stand for election to be the president of your learned society. It will be a valuable contribution if you get the strategy right and also if you assist the colleagues elected with you to perform well. You will also have the opportunity to make many new friends and meet many existing ones on a regular basis.

30 How to Lead Your Research Community as an Instrument Scientist

Data are the means by which scientific hypotheses are rejected or confirmed and are the platform on which the words of publications firmly rest. An instrument scientist therefore plays various important roles in experimental science and the data analyses that flow from it. A great deal of my career has involved facilitating the best measurements with X-rays using X-ray diffraction as a tool, mainly not only from crystal samples but also from fibres, solutions and, latterly, from crystalline powders, largely, but not entirely, involving proteins. In the last 20 years or so, I have also been involved in securing the best possible measurements using neutrons in protein crystallography, but not directly as an instrument scientist for those measurements myself; those have been made in collaboration with my former PhD student, Dr Matthew Blakeley, Institut Laue–Langevin Neutron Instrument Scientist for the Laue diffraction diffractometer.

The term *instrument scientist* means the scientist who ensures a well-functioning instrument, the best data measurements possible and a secure processing of the primary (raw) experimental data. The handover of these processed data to the research group who devised the experimental proposal in the first place can be made and whose scientists can derive further results, and which we call the derived data. Of course, these respective roles of the members of a research project team do not exactly follow these simplified definitions, but they are a reasonable, albeit approximate, description of the scientific roles played. It is also to be encouraged that the instrument scientist forms his or her own scientific research projects and carries through the project to results, discussion and conclusions. I was for seven years an instrument scientist at the UK's SRS where I led the design, the commissioning and the utilisation by the UK and the international research user community of two instruments. These were the instruments SRS 7.2 and 9.6 (see Figure 30.1 [1,2]). At the SRS, the management terminology was that we were called *station scientists* or *beamline scientists*. Those terms were fine. Much later, when I took up the role as director, on secondment from my academic professorship at the University of Manchester, the terminology had changed from *Station scientist* to *station manager*, a label that I really disliked as the role was much more than *manager*.

Through my career, I have served on various beamtime application panels judging proposals requesting beamtime on instruments of the type I had myself developed at the SRS; for example, at the Italian Elettra synchrotron, the Spanish ALBA synchrotron and later also on the Institut Laue–Langevin (Europe's research reactor-based neutron source) chairing their biology with neutrons subcommittee 8. Currently,

FIGURE 30.1 Here I am in the SRS 7.2 hutch; actually, this photo was taken in 2002 when I had recently taken up the role as director of SRS at the UK's Council for the Central Laboratories of the Research Councils. (From Science and Technology Facilities Council and Stuart Eyres. With permission.)

I am chair of the European Spallation Source Instrument Science and Technical Advisory Panel for neutron macromolecular crystallography and a member of the Instrument Advisory Panel for the long wavelength macromolecular crystallography beamline at the UK's third generation SR X-ray source (a synchrotron X-ray source of hugely improved technical specification and performance over the SRS).

The best of instruments and the best of instrument scientists attract beamtime proposals from the best research groups and even serve as a magnet pulling research groups away from their home countries. With both SRS 7.2 and SRS 9.6, we attracted leading scientists from the United States and Sweden, for example, and they undertook pioneering research. In my Patterson and Perutz Prize Awards, respectively, that I received from the American Crystallographic Association in 2014 and the European Crystallographic Association in 2015, the award citations included statements about how I had trained some of the best beamline scientists in the world; I am also a proud scientific father indeed!

The important roles of the instrument scientist are described in a really nice article by Dr Estelle Mossou in the October 2015 *Physics World* magazine (Figure 30.2). She explained how she balances the demands of users while carrying out her own research [3] and from which I quote (with permission of the author, the Institut Laue–Langevin and *Physics World*):

> The choice of instrument and technique very much depends on the problem at hand – academic or industrial. As an instrument scientist, my tasks are diverse and variable. The primary one is to provide support to users for a variety of scientific experiments. Here my job is to assist the experimental team to obtain the best possible results from samples the preparation of which has often taken many months of work. The planning of the experiment will strongly depend on the specific nature of the sample, the length of beam time allocated to that particular experiment and obviously what the users

FIGURE 30.2 Dr Estelle Mossou, Institut Laue–Langevin Neutron instrument scientist for the D19 diffractometer. (Courtesy of Institut Laue Langevin, Serge Claisse; from Dr Estelle Mossou, the Institut Laue Langevin and *Physics World*. With permission.)

need and expect. The instrument-scientist role also implies a strong involvement in the development and upgrade of the instrument – something that is carried out very much in consultation with the external user community. Indeed, D19 was recently rebuilt and upgraded thanks to a grant from the UK's Engineering and Physical Sciences Research Council.

One aspect that will perhaps cause the most diverse of behaviours from your users will be publications. Facilities today generally have a fairly strict rule concerning the involvement of the instrument scientist in publication, with an expectation of coauthorship. In my years as a station scientist, there was no formal management view. At the time, I was contacted by colleagues in other facilities to introduce an obligatory rule of coauthorship for my users of SRS 7.2 and 9.6. I resisted this. After all, I said, what if two users got into a competition? I would be left in an impossible situation. Instead, I commended that a publication would be written about one's instrument and users would be required to cite it, as a minimum. Through my years, there were, however, situations where I really did think that coauthorship was appropriate, as the work together with a user group then did feel collaborative. But there was nothing written down in advance. I just had to grin and bear it. Much later, when I was a professor, a situation arose where I was consulted about a case involving a research officer who became aware of an article submission only when a postdoc came along asking for help to deal with referees' queries; the research officer not being a coauthor! This was clearly a case where the research officer should have been a coauthor, who told me that 80% of what was in the paper was material and results the research officer had provided. These days, journals require a corresponding author to declare the roles of each author, and it is exactly what is required to combat such scientific malpractice. To have clarity from the facility or the university management in advance regarding coauthorship is, in retrospect, I think better than my own attempt to steer a course on such a potentially thorny issue of publication expectations from users. Basically, I think it is quite fair that instrument scientists who have made a significant contribution to the concept, the design, the execution, the analysis or the interpretation of user experiments should be offered the opportunity to be listed as coauthors in publications.

As examples of the leadership roles the instrument scientist can play, I would mention the following:

Firstly, you get an incredible overview of your whole research field based on which you can spring forward with significant initiatives for your field of research.

Secondly, you have a chance with your instrument to surpass all others in the same instrument class; you can make your instrument sing its song by ensuring the best possible design and that it works well. In those years that I was instrument scientist at SRS with 7.2 and 9.6, I had to work with a horizontal source size of 14 mm, but the samples were always less than 0.5 mm! But still, these two instruments sang their important song and attracted the best researchers in the world for many years.

Thirdly, you naturally have a strong interest in the data analysis software. You may be good at writing software yourself. I did some of that, but others, notably graduates in computer science, were better at software code writing than myself. Anyway, I instigated and coordinated a round robin study where one set of diffraction data were processed with different software and by different research groups. This had two objectives: to know which were the best algorithms for analysis of data and also to ensure a good research practice by the research groups who were my instrument users. I presented the results in a lecture at the World Congress of Crystallography in Ottawa, in 1981, my first major lecture in fact [4].

Why did I leave behind my roles as station scientist for SRS 7.2 and 9.6? I was suffering from severe exhaustion from long hours of working, often missing a night's sleep and still trying to conduct my day job. I was also trying to undertake methods development research as well as user program local contact support. When the opportunity arose, I moved into a university joint appointment with the SRS and which style of my employment continued for many years, only finally ceasing in 2007 (the SRS closed in 2008), 31 years after my first experiment at Daresbury, on the NINA synchrotron, which preceded the SRS and for which the SR was a parasitic use of the predominantly high-energy physics machine. I describe the early years (1976–1993) of my scientific work at Daresbury in Reference 5.

Overall, the instrument scientist can have a profound role leading important change in a research field and initiating new research fields.

REFERENCES

1. J. R. Helliwell, T. J. Greenhough, P. Carr, S. A. Rule, P. R. Moore, A. W. Thompson and J. S. Worgan (1982) Central data collection facility for protein crystallography, small angle diffraction and scattering at the Daresbury SRS. *Journal of Physics E* 15: 1363–1372.
2. J. R. Helliwell, M. Z. Papiz, I. D. Glover, J. Habash, A. W. Thompson, P. R. Moore, N. Harris, D. Croft and E. Pantos (1986) The Wiggler protein crystallography work-station at the Daresbury SRS: Progress and results. *Nuclear Instruments and Methods* A246: 617–623.
3. E. Mossou (2015) Life as an Instrument Scientist: Physics World October 2015 Focus on Neutron Science. Accessed 7 August 2016. Retrieved from http://live.iop-pp01.agh.sleek .net/physicsworld/reader/#!edition/editions_neutron-2015/article/page-9655.

4. J. R. Helliwell, A. Achari, A. C. Bloomer, P. E. Bourne, P. Carr, G. A. Clegg, R. Cooper et al. (1981) Protein crystal oscillation film data processing: A comparative study. *Acta Crystallographica* A37: C311–C312.
5. J. R. Helliwell (2015) Protein crystallography at Daresbury: The early years (1976 to 1993). Accessed 7 August 2016. Retrieved from http://www.synchrotron.org.uk/index.php ?option=com_content&view=article&id=67:srs-protein-crystallography-the-early -years-1976-to-1993-by-john-r-helliwell&catid=39:bio-science&Itemid=53.

Section VII

Ethics, Global Development, Policy and the Organisation of Science

31 How to Retain Your Own Peace of Mind
The Ethical Aspects

Integrity without knowledge is weak and useless, and knowledge without integrity is dangerous and dreadful.

Samuel Johnson
The History of Rasselas, Prince of Abyssinia (now Ethiopia)

I have something that I call my Golden Rule. It goes something like this: Do unto others twenty-five percent better than you expect them to do unto you. … The twenty-five percent is for error.

Linus Pauling
*His reply to an audience question about his ethical system,
following his lecture circa 1961 at Monterey Peninsula College, California*

There are two aspects to ethics in science I wish to deal with. Firstly, there are the ethical implications of (some) scientific discoveries. Secondly, there are the ethics of how we respect other scientists.

Perhaps the most well-known example of scientists wrestling with the ethical aspects of a discovery is the physics behind the critical mass of uranium needed to sustain a chain reaction and thereby a nuclear explosion. The counterpoint to this new science implication was the knowledge that this was also a new source of controlled production of our energy supplies. On the perils of Germany developing the first atomic bomb, Einstein signed a letter drafted by Szilard to the then president of the United States, Franklin D. Roosevelt, on 2 August 1939, commending that research be undertaken to enable this to be achieved first by the Allies, rather than the Nazis. Einstein later expressed regret that he had signed the letter. In response to the Einstein–Szilard letter to him, President Roosevelt established an advisory committee, and whose recommendations developed into the Manhattan project to develop such a weapon led by J. Robert Oppenheimer (1904–1967) who was

an American theoretical physicist and professor of physics at the University of California, Berkeley. As the wartime head of the Los Alamos Laboratory, Oppenheimer is among those who are called the 'father of the atomic bomb' for their role in the Manhattan Project, the World War II project that developed the first nuclear weapons that ended the war with the atomic bombings of Hiroshima and Nagasaki. The first atomic bomb was detonated on July 16, 1945, in the Trinity test in New Mexico; Oppenheimer remarked later that it brought to mind words from the Bhagavad Gita: 'Now I am become Death, the destroyer of worlds'.

After the war Oppenheimer became chairman of the influential General Advisory Committee of the newly created United States Atomic Energy Commission, and used that position to lobby for international control of nuclear power to avert nuclear proliferation and a nuclear arms race with the Soviet Union [1].

In a totally different field of science, genome editing, Professor Jennifer Doudna recently wrote [2] about the ethical dimensions of her gene editing science and the potential it has, for example, for editing of human embryo genes, including her ethical concerns (quoted with the permission of Professor Doudna and from Macmillan Publishers Ltd):

> I am excited about the potential for genome engineering to have a positive impact on human life, and on our basic understanding of biological systems. Colleagues continue to e-mail me regularly about their work using CRISPR–Cas9 in different organisms – whether they are trying to create pest-resistant lettuce, fungal strains that have reduced pathogenicity or all sorts of human cell modifications that could one day eliminate diseases such as muscular dystrophy, cystic fibrosis or sickle-cell anaemia.

Professor Doudna also champions the need for scientists to engage with outreach to the public on the ethical consequences of their work (quoted with the permission of Professor Doudna and from Macmillan Publishers Ltd):

> But I also think that today's scientists could be better prepared to think about and shape the societal, ethical and ecological consequences of their work. Providing biology students with some training about how to discuss science with non-scientists – an education that I have never formally been given – could be transformative. At the very least, it would make future researchers feel better equipped for the task. Knowing how to craft a compelling 'elevator pitch' to describe a study's aims or how to gauge the motives of reporters and ensure that they convey accurate information in a news story could prove enormously valuable at some unexpected point in every researcher's life.

I was glad to meet Jennifer Doudna and enjoy dinner with her at the Argonne National Laboratory Guest House Restaurant in around 2005 when she had joined the Advanced Photon Source Science Advisory Committee.

Not many scientists confront so vividly such dramatic implications of their research as the two examples I have described above. However, for us all to study such examples as case studies is important. Also, as Professor Doudna champions, it is vital for us as scientists to be well prepared to face ethical questions, and how to do this should surely be a mandatory part of the skills training we receive. One thing is for sure, the discussions of the implications of scientific research discoveries will not be for us to define on our own, such discussions must include all constituents of society and, at the least, the society's elected representatives. Conversely, these elected representatives must include scientists in such debates to provide firm contact with the scientific facts. I also think that the scientists should be there on an equal footing, with voting rights, just the same as the politicians. This latter point is not necessarily accepted by politicians! Most famously, Winston Churchill remarked, 'Scientists should be on tap but not on top'.

On the second aspect of ethics, namely of how we respect other scientists, a source book of ideas that are quite fundamental about people, good and bad, is by Erich Fromm, philosopher and psychologist, *Man for Himself: A Psychology of Ethics* (1947) [3]. A few quotations from his book indicate the good that can come from being, well, good:

Man's main task in life is to give birth to himself, to become what he potentially is. The most important product of his effort is his own personality.

Then Fromm takes the well-known saying 'Whatever you do to others, you also do to yourself'. And inverts it to

Do not do to others what you would not have them do to you.

On good behaviour to others, he makes the insightful remark:

No healthy person can help admiring, and being affected by, manifestations of decency, love and courage, for these are the forces on which his (one's) own life rests.

Fromm explains in detail the obverse case:

(there are people) with an exploitative personality. . . . They use and exploit anybody and anything from whom or from which they can squeeze something. And furthermore that such people want to use and exploit people, they 'love' those who, explicitly or implicitly, are promising objects of exploitation, and get 'fed up' with persons whom they have squeezed out.

Fromm deals with the nurture versus nature aspects of the types of people mentioned earlier by separating temperament, namely what is in our genes, from character, namely what we make of ourselves. He also emphasises that people are in general a blend of good and bad, rather than extremes.

In the end, Fromm observes that 'the decision of what man makes of himself is man's'. Finally, Fromm cites the famous quotation of Plato which addresses the place of individuals within their external environment, namely society itself [4]:

Until philosophers are kings, or the kings and princes of this world have the spirit and power of philosophy, and political greatness and wisdom meet in one, and those commoner natures who pursue either to the exclusion of the other are compelled to stand aside, cities will never have rest from their evils, no, nor the human race, as I believe – and then only will this our State have a possibility of life and behold the light of day.

The psychology analyses of people by Fromm that I have described above are quite general. Let's now move on to specifically discuss the category of scientists. In 2007, the UK government's chief scientific advisor, Professor Sir David King [5], laid out a universal code of ethics for scientific researchers across the globe. The UK government has apparently adopted them, although it is not clear quite what that means in practice!

David King's seven principles of the universal code of ethics, intended to guide scientists' actions, are as follows:

1. Act with skill and care in all scientific work. Maintain up to date skills and assist their development in others.
2. Take steps to prevent corrupt practices and professional misconduct. Declare conflicts of interest.
3. Be alert to the ways in which research derives from and affects the work of other people, and respect the rights and reputations of others.
4. Ensure that your work is lawful and justified.
5. Minimise and justify any adverse effect your work may have on people, animals and the natural environment.
6. Seek to discuss the issues that science raises for society. Listen to the aspirations and concerns of others.
7. Do not knowingly mislead, or allow others to be misled, about scientific matters. Present and review scientific evidence, theory or interpretation honestly and accurately.

The details mentioned earlier involving people with whom you work with or meet clearly relate to other chapters in this book such as Chapter 6 (How to Set Up, Lead and Care for Your Research Team), Chapter 13 (How to Coexist with Competitors) and Chapter 15 (How, and When, to Effect Collaborations). However, so much of what you yourself do, and how you conduct yourself, relates to this chapter!

The code of principles covers good practice. Let us also consider the obverse and ask what, formally, research malpractice is. As an elected representative of the Faculty of Science and Engineering to the Senate of the University of Manchester, I took part in the discussion, the amendments and the final approval of the university's policy on research malpractice. This important document [6] had as core elements the definition of misconduct in research, the categories of poor research practice and the categories of research misconduct. The policy of the university covered many types of research both in detail and overall. While I have not been involved in experiments on animals or humans, the points relating to such experiments looked to me reasonable, as they were also to the medical academics on the university. One area that particularly caught my eye, since I have a heavy reliance in my crystallographic research on experimental raw data, processed data and final derived data, was the category of research misconduct relating to data namely 'Mismanagement or inadequate preservation of data and/or primary materials'. While crystallographers, particle physicists, astronomers, climate change specialists and genome sequence researchers, for example, are well known for their efforts to preserve, and ensure access to, their data, I was aware of less good practice more generally, but that is certainly changing. In the United Kingdom, this is, for example, through the fine efforts of the Digital Curation Centre and internationally through the International Council for Science's CODATA and the Research Data Alliance. Locally, the University of Manchester now ensures preservation of data and publications by its researchers, via its eScholar* web tool administered by the university library.

---

* eScholar at the University of Manchester is being updated to the new system PURE.

One of the very specific areas to which ethics applies involves publications. Very generally, as Wikipedia states, publication ethics is 'the set of principles that guide the writing and publishing process for all professional publications'. Most obviously, it should be ensured that a publication is not plagiarised in whole or in part. This covers not only plagiarism of another's work but also self-plagiarism, namely the reuse of text, figures, tables or your data in more than one of your publications. The area of publication ethics is, however, more multifaceted than plagiarism, and cases can be quite complex. An organisation called the COPE exists [7], which provides 'a forum for editors and publishers of peer reviewed journals to discuss all aspects of publication ethics. It also advises editors on how to handle cases of research and publication misconduct'. A resource that COPE provides is a whole suite of cases [8]; at the time of writing, there were 531 cases listed! COPE also assists the readers of their case studies in providing a case taxonomy. At the time of writing, this comprises 18 main classification categories and 100 keywords. These categories, slightly expanded by me for clarity and or inclusiveness of relevant topics, are as follows:

Authorship; Conflicts of interest; Consent for publication; Contributorship; Copyright disputes/breaches; Correction of the literature; Data; Editorial independence; Funding/sponsorship; Impact Factor manipulation and metrics; Books, social media and legal issues; Misconduct/questionable behaviour by the author(s), by a journal, by an editor, by a publisher, by an institution or by a reviewer; Mistakes; Peer review and/or editorial decisions; Plagiarism; Questionable or unethical research; Redundant/duplicate publication or multiple submissions; Responding to whistle blowers.

As an example of the perspectives of a publisher, the International Union of Crystallography, for which I served nine years as Editor-in-Chief, its ethical publishing policy is described in Reference 9. Within this policy is the practical tool that can be deployed by a publisher to help publishers screen articles for plagiarism [10].

As a second publisher perspective, I mention Taylor & Francis, as I have served for 10 years as a main editor of one of their journals, *Crystallography Reviews*, and for the last 3 years, as the sole main editor. Their publication ethics policy is described in Reference 11. They also offer an infographic for authors (Figure 31.1) and a useful checklist, which I reproduce below, also with Taylor & Francis' permission:

"Ready to submit your paper? Your ethics checklist
    Before you submit, make sure:

   i. You've read the journal's instructions for authors, and checked and followed any instructions regarding data sets, ethics approval, or statements.
   ii. All authors have been named on the paper, and the online submission form.
   iii. All material has been referenced in the text clearly and thoroughly.
   iv. Data have been carefully checked and any supplemental data required by the journal included.
   v. Any relevant interests have been declared to the journal.
   vi. You've obtained (written) permission to reuse any figures, tables, and data sets.
   vii. You've only submitted the paper to one journal at a time.
   viii. You've notified all the co-authors that the paper has been submitted."

FIGURE 31.1 Taylor & Francis's ethics for authors' guidance before submission of an article [11]. (Reproduced with the permission of Taylor & Francis.)

One aspect that editors, referees and journals are not so successful at monitoring is the comprehensiveness of the citations of relevant references in a research article. Authors, if queried, usually in my experience respond well to suggestions in referees' reports to cite new references and/or state why the suggested new references are not relevant. But this is a haphazard business in my view, and citation tools could surely be harnessed to assist the editor and the referees to ensure a better coverage of relevant literature. As an objective measure of the problem, I will mention the journal citation half-life metric. The cited half-life is the number of publication years accounting for half of the citations received. Formally, this is portrayed as 'an indicator of the turnover rate for a body of work on a subject', and partly, this is no doubt true, but I think it also reflects the memory span of the authors, the referees and the editors and thereby what goes into the article reference lists!

While publications or publication practices that are unethical get exposed sooner or later, a different area of our scientific life is that of research grant proposals, which involves much less transparent procedures. By this I mean that the proposal is unlikely to be openly published, although increasingly a summary of what has been funded is now often available. Here a proposal that is turned down for good reasons, or the quite usual modern situation, 'insufficient funds albeit of international quality', is vulnerable to malpractice. What do I mean by this? A funding agency must turn to experts for referee reports, and in turn, experts may well be, sooner or later, competitors. I have had a situation where it did seem a remarkable coincidence to me to witness a publication on the very area that I had previously proposed coming from a lab where it seemed, in retrospect, that the lab would also have likely been a referee of my proposal. As Fromm observed, there are exploitative people out there. Nobel Prize Winner Bill Lipscomb in the *New York Times*, quoted by Henry Bauer [12], stated that '(I) no longer put my most original ideas in my research proposals, which are read by many referees and officials. I hold back anything that another investigator might hop on and carry out. When I was starting out, people respected each other's research more than they do today, and there was less stealing of ideas'. Sobering thoughts! But would Bill Lipscomb's approach work today? Lipscomb had a Nobel Prize then, had all the prestige in the world, could write a research grant on basically anything and get funded (and remember that those were the days of 30–40% research grant success rates anyway versus today of around 10–15%). So his model would simply not work in today's research world. With research grant success rates at 10% as a principal investigator (PI), one has to place the best foot forward and reveal one's ideas and plans every time if you want your lab to be funded.

I recently attended a lecture by someone whose research in chemistry had strayed into an area, he said, where it was well known as being done by 'an ocean of sharks'! Afterwards, I asked the lecturer what skills or tactics he had deployed in handling the ocean of sharks to which he had referred?! He said he had been lucky as it was not a research grant proposal, rather an unfunded off shoot of another work in his lab. In the publication, his emphasis was to avoid the main keywords that would have suggested to a journal editor to approach one of the sharks. The sharks were thereby avoided! Overall, probably the way to deal with 'sharks' is to have enough ideas, and enough active research, so that if you get bitten, you can still swim.

In summary, the descriptions mentioned earlier give the necessary guidelines, and indeed laws of the land, for a scientific researcher from my perspective at a leading research university and of two publishers of science that I know well. In addition, my remarks provide a wider context, including my quoting the point of view of a psychologist, Erich Fromm, with an emphasis on how it can go wrong in part or in whole! In any field of scientific research, there will be some degree of problems of fabricated data and results as evidenced by withdrawal of publication by a journal, such cases being often without the authors' consent, and who seemingly have just disappeared! Fortunately, such cases in any field and across the whole of science are very rare. But let us now ask what success would look like in a truly ethical scientific landscape. For this, we must of necessity look to highly regarded individuals. I leave it to you the reader to identify those individuals who in your impression in your field of research behave ethically. As measures, you can check, for example:

1. Do they cite their own lab members carefully in their talks?
2. Do their publications cite the relevant scientific literature, which you will know well the more so as your career progresses and your experience deepens?
3. Do their publications include the relevant data to substantiate their words?

No person is an island, and it is on this simple but important foundation that our ethics in my opinion basically rests. You must therefore develop your feeling and sensitivity to others, while respecting of course your own dreams and visions for your science. You should also not be afraid to assert your views if someone treats you unethically, but if so, you will need to proceed with tact and with clarity of your evidence in your mind.

Organisations and indeed even possibly your employer may act unethically. Receipt of funds from dubious individuals or companies seeking special treatment at your institution are the most obvious cases to voice your opposition about. But internally to an organisation, there are, to me, some strange behaviours that can manifest: For instance, how would you view the writing of documents by staff for their line manager, whose name alone then appears on that document? I view the matter with dissatisfaction. Secondly, how would you view a facility or a research proposal from one laboratory, which is approved, but then for other reasons, for example, regional geography, is forced to be undertaken elsewhere? In the end, you will have to be true to yourself, and those that depend on you, to decide how to proceed on such matters, which can be exceedingly awkward in practice, although clear in the principles.

A book that presents real life case studies of ethical violations in the history of science, whilst providing a firm basis for discussing them, is by D'Angelo [13]. As your own overall guide, I commend that you trust your conscience and by which means you should be able to retain your own peace of mind in your work and your life.

REFERENCES

1. Wikipedia (n.d.) J. Robert Oppenheimer. Accessed 7 August 2016. Retrieved from https://en.wikipedia.org/wiki/J._Robert_Oppenheimer.

2. J. Doudna (2015, December 24) Genome-editing revolution: My whirlwind year with CRISPR. *Nature* 528: 469–471.

3. E. Fromm (1947) *Man for Himself: A Psychology of Ethics.* Routledge, New York and Kegan Paul Ltd, London: pp. 64, 65, 111, 245.

4. Plato (380 BC) *The Republic.* In Great Ideas of Western Man series. Oxford University Press, Oxford.

5. D. King (2007) A Hippocratic oath for scientists. Accessed 25 November 2015. Retrieved from http://blogs.nature.com/news/2007/09/hippocratic_oath_for_scientist.html.

6. The University of Manchester *Policy on Research Practice.* Accessed 7 August 2016. Retrieved from http://documents.manchester.ac.uk/display.aspx?DocID=611.

7. Committee on Publication Ethics (n.d.) Retrieved from http://publicationethics.org/.

8. Committee on Publication Ethics (n.d.) Case Studies. Retrieved from http://publicationethics.org/cases.

9. International Union of Crystallography *Journals' ethical publishing policy.* Accessed 7 August 2016. Retrieved from http://journals.iucr.org/services/ethicalpublishingpolicy.html.

10. CrossRef (n.d.) Similarity check. Retrieved from http://www.crossref.org/crosscheck/index.html.

11. Taylor & Francis's *Publication ethics policy.* Accessed 7 August 2016. Retrieved from http://authorservices.taylorandfrancis.com/ethics-for-authors/.

12. H. H. Bauer (1989) Nobel fever: The price of rivalry (1989, October 17), *New York Times*: pp. C1,14. In *Ethics in Science.* Accessed 7 August 2016. Retrieved from http://www.umsl.edu/~chickosj/202/Ethics/Ethics%20in%20Science.pdf.

13. J D'Angelo (2012) Ethics in Science: Ethical Misconduct in Scientific Research CRC Press, Boca Raton, Florida.

32 How to Make Your Role a Global One

Your role as scientist can extend far and wide.

Your lab can host students and postdocs from other countries besides your own. Your participation in international conferences is a core activity for any scientist, learning of others' discoveries and presenting your own. As you get more experienced, you will have the chance to actively help organise conferences and a core role in ensuring the International Council for Science's free circulation of scientists principle [1] through careful approval of the choice of hosting country; this obviously at least means the promise from a hosting country to provide travel visas to conference participants in a timely manner for anyone wishing to attend. This very important principle is worth quoting from the ICSU* website in full as it impinges on a variety of other practical aspects [1]:

ICSU Statute 5

The Principle of Universality (freedom and responsibility) of Science: the free and responsible practice of science is fundamental to scientific advancement and human and environmental well-being. Such practice, in all its aspects, requires freedom of movement, association, expression and communication for scientists, as well as equitable access to data, information, and other resources for research. It requires responsibility at all levels to carry out and communicate scientific work with integrity, respect, fairness, trustworthiness, and transparency, recognising its benefits and possible harms.

In advocating the free and responsible practice of science, ICSU promotes equitable opportunities for access to science and its benefits, and opposes discrimination based on such factors as ethnic origin, religion, citizenship, language, political or other opinion, sex, gender identity, sexual orientation, disability, or age.

N.B. This wording of Statute 5 has been approved by the 30th ICSU General Assembly in Rome in September 2011.

The part of the statute 'equitable access to data, information, and other resources for research' presents the need to help with capacity building around the world wherever it is needed. Of course, up-to-date facilities and resources are needed everywhere! Some of the most pressing cases for capacity building of course are in the developing world, where you can help by transferring your hard-won skills by your taking part in local training workshops, in training workshops before and after conferences, as well as in supporting bursary applications for travel by students to these workshops and to the major conferences. Your lab, as it grows, is likely to be multinational, one way or another. My lab hosted many nationalities over the last 40 years.

---

* International Council of Scientific Unions (ICSU) is now called the International Council for Science.

But how many of your PhD students and staff returned to their home country? Very few? If so, then in essence, they became part of their country's brain drain; I mention it that way as an encouragement that you need to do more than have a multinational lab if you wish to help facilitate globalisation and, as I say, transfer your skills directly to other countries, which I hope you will do. These points deal with equitable access to resources for research and for training but what about equitable access to data and information?

Equitable access to data and information necessarily means access to the publications and the data that underpin those publications. ICSU convened a workshop on these matters and produced a report [2], endorsed at the ICSU 31st general assembly held in Auckland, New Zealand, in 2014. I represented the International Council of Scientific and Technical Information (ICSTI) at the workshop. The report emphasised the following goals:

> The International Council for Science advocates the following goals for open access. The scientific record should be:
>
> i. free of financial barriers for any researcher to contribute to;
> ii. free of financial barriers for any user to access immediately on publication;
> iii. made available without restriction on reuse for any purpose, subject to proper attribution;
> iv. quality assured and published in a timely manner; and
> v. archived and made available in perpetuity.
>
> These goals apply both to peer-reviewed research publications, the data on which the results and conclusions of this research are based, and any software or code used in the course of the research.

Furthermore, in the world of research, the ICSU report also correctly stated that

> The process of scientific discovery involves researchers being able to communicate their research results and audiences being able to access these results. At the same time, not all researchers have ready access to research funding and cost must not be a barrier to placing their results in the most appropriate journal, data repository or other outlet for that research.

It can be estimated that funded research comprises a minority of the research that is undertaken. Therefore, finding sustainable ways to maximise the science that is communicated by authors and to maximise the science that is accessed by readers is a paramount concern. You can influence this by ensuring that at least some of your research is open access to readers, ideally all of it. This ideal is difficult to realise in my experience because, like most lab leaders, only a fraction of my research grant proposals got funded; the precise fraction varied over the years but was typically around 25%. The ideas and the proposals that did not get funded, the majority, were still in my view excellent ideas and proposals, and I would take them forward somehow, usually in a reduced form and over a longer period. The work would be done by me and/or with final year undergraduate project students. This was good in the sense that my own skills had to be kept up to the mark, which I enjoyed. But, most likely, I could not afford the author article processing charges to publish those results in open access to the reader journals. I have been glad of reader subscribers paying to read my research, even if it did mean those publications are behind a paywall.

In my research field of crystallography, there are, quite rightly, stringent requirements that articles will only be published with the accompanying data. Thus, in the crystallography life sciences, which is one of my research specialisms, Protein Data Bank (PDB) deposition files (derived coordinates and processed structure factor amplitudes) are obligatory to accompany a publication. The PDB is a publicly funded database, and so access to depositors is free of any submission charge as well as to its users. Let us hope this continues! I believe such important databases should be free to deposit, as, overall, the data provide the bedrock on which our publications rest. They allow readers to repeat many of the calculations that they wish in addition to reading the words in a publication. To check all the calculations, readers need access to raw data. This is all a necessary prerequisite to contributing to a global sustainable future accurately and precisely based on a full scientific record including data and information. You should take as much care over your data deposits as your honing of the best, clearest possible written publication. With that said, much remains to be done to improve the refereeing of articles with data; it is a challenge to ask a referee to consider both, and it has been patchy (some journals do, and many other journals do not), but I have lobbied wherever possible for this. In the end, it may require journals, and databases, paying referees to do this, but without the refereeing of both the written word and the associated data, there will be the danger of inconsistency of publication and data standards. The amount of work to be undertaken by a referee in the future is also about to further increase as the archiving is extending further down the data pyramid to the raw experimental data sets, not only the processed and final derived data, which are smaller file sizes than the raw data. Anyway, overall, as you get more senior, you can also accept representative roles in science, as I have done, applying your hard-won skills and experience and join these important debates at the highest levels of global scientific life.

REFERENCES

1. International Council for Science (2011) ICSU Statute 5. Accessed 7 August 2016. Retrieved from http://www.icsu.org/freedom-responsibility/cfrs/statute-5.
2. International Council for Science (2014) Open access to scientific data and literature and the assessment of research by metrics. Accessed 7 August 2016. Retrieved from http://www.icsu.org/general-assembly/news/ICSU%20Report%20on%20Open%20 Access.pdf.

33 How to Make Your Input into Science Policy

I believe it worthwhile to express one's views when there is a call for evidence by one's elected representatives. I present my inputs to the UK Parliament, which had calls for evidence on the very general issues of interdisciplinary science and on publication of scientific research.

In 1999, I wrote on the subject of interdisciplinary science research funding to the UK Parliament's Select Committee on Science and Technology, which published my input in their Appendices to the Minutes of Evidence [1]. Here is an extract with the main point that I made:

> I wish to raise the issue of the present state of organisation of UK scientific research funding (by subject specific Government research councils) at the interdisciplinary science interface, namely the junction between biology, chemistry and physics as encapsulated by my research. The challenges met in society today with issues of health and agriculture rest on scientific methodology and results which simply do not respect the tight subject boundaries of our subject specific UK Research Councils and require a different approach for interdisciplinary science research.

I took up the theme of interdisciplinary science again in *Nature* in 2007 [2], of which this is a short extract of my article:

> The letter from Smith and Carey (*Nature* 447, 638–639) [3] addressed the need for supportive environments if interdisciplinary research goals are to be achieved. Specifically they highlight the lack of institutional structure and support for such research. Also, they mention that assessments of such research quality are made in single subject committees in Australia, New Zealand and the UK. I sympathise with their point of view and took my concerns to the UK Parliamentary Science and Technology Committee general review of UK research in 1999 (http://www.parliament.the-stationery-office .co.uk/pa/cm199900/cmselect/cmsctech/196/196ap02.htm).... Obviously my wish for a unified science and engineering research council, with a single charter and unified management structure, was not achieved of course but I feel that official efforts were made to respond to my criticisms and have put the UK in a more flexible position by my having 'spoken out' about interdisciplinary science funding.

See also the summary input from the associate editor of *Nature*, Maxine Clark [4].

I also wrote to the UK's House of Lords Parliamentary Select Committee on Science and Technology's consultation on open access for research publications [5], of which this is an extract of my submission to their Lordships:

> How can science communication publication survive in an author-pays era? I especially address the needs of authors who cannot afford Gold Open Access fees for the

'quality assured publications' method of science communication … which will commence for RCUK funded research from April 2013…. Indeed what is the alternative for communication of the non-funded research results?

I return to this last question in Chapter 34.

Overall, one does not know if one's input has a direct impact, as there is no specific feedback offered to those making such inputs. At the heart of it though, at least one has done one's duty and made one's expertise and experience available to one's elected representatives.

REFERENCES

1. UK Parliament's Select Committee on Science and Technology Appendices to the Minutes of Evidence (1999) Retrieved from http://www.parliament.the-stationery-office.co.uk/pa/cm199900/cmselect/cmsctech/196/196ap02.htm.
2. J. R. Helliwell (2007, August 2) Correspondence: Interdisciplinary research could pull cash into science. *Nature* 448: 533. Retrieved from http://www.nature.com/nature/journal/v448/n7153/full/448533b.html.
3. J. A. Smith and G. E. Carey (2007) Correspondence: Those who are crossing boundaries need less talk, and more help and flexibility. *Nature* 447: 638–639.
4. M. Clark (2007) Creating an interdisciplinary research culture. Accessed 7 August 2016. Retrieved from http://blogs.nature.com/nautilus/2007/06/creating_an_interdisciplinary.html.
5. Report of the UK's House of Lords Parliamentary Select Committee on Science and Technology's consultation on open access for research publications (2012–2013) Accessed 7 August 2016. Retrieved from http://www.publications.parliament.uk/pa/ld201213/ldselect/ldsctech/122/122.pdf.

34 How Would You Change the Organisation of Global Science If You Were in Charge for One Day

This is your chance to make a wish for improving the world's science! I would pick the following wishes:

> 1. I propose an open access publications fund for unfunded research.

This would be relevant to those grant proposals that the world's funding agencies judged were fundable in principle, for example, with grades that were outstanding or internationally excellent, but left unfunded because an agency sadly had no more research funds available. In my experience as principal investigator (PI), I try and find some way to take the research forward, for example, in some restricted form financially and in a longer time than proposed. Finally, at the time of publication, the original funding agency could be recontacted by the PI, and the funding agency would be invited to pay the open access fees of the publication from a special fund that would be set aside for the purpose. My point being that open access for one's publications is great, but to do that, you need the monies to pay for it, and for unfunded research, that is likely to be problematic. While the funding agencies contemplate that, then we can use those preprint servers that can provide free open access publishing with a digital object identifier (DOI). Such preprint servers are not a traditional journal with the usual peer review but provide a home for a real, citable open access paper.

> 2. I propose that all submitted science research articles be accompanied with their data.

The quality of the research publications, and the database deposits that should accompany a published article, would thereby improve, as the referees of course would have access to the data as well as the words of the article; this would be obligatory because the publication would then de facto be words and data. Moreover, the reader of the article, and thereby the potential user of the data accompanying the article, would not have to take the word of the authors for it; the reader/user could repeat the calculations for themselves. The motto of every journal should be 'Do not just take the authors' word for it'. In the past, there has been a possible argument of mitigation that the primary experimental data sets were too large to archive, and only the processed data and the final derived data from that processed data could be archived. With the marvellous improvement of computer disc storage capacities and centralised data stores, even commercially available, the need to delete/lose/not keep the raw data for a given experiment is now very often not necessary. Moreover, the registration of a data set with its own DOI, and clear guidance for citation of data sets through the leadership of the ICSU's CODATA, means that a data set is as much an important piece of knowledge as a publication's words. Also, repeat analyses with a data set may well be with improved software algorithms compared to the software used at the original date of the analysis and the publication, and so our scientific culture should encourage, indeed require, all submitted science research articles be accompanied with their data.

  3. I propose that all scientists to give some of their time towards world peace and sustainability.

I wish for all scientists to take part in tackling world peace and sustainability. There are many ways for you to make a contribution. In my own research field, two prominent, capacity-building cooperative projects include the Middle East Synchrotron and the proposed African Synchrotron Light Source. These new facilities illustrate the importance of a sufficiently peaceful world and environment for research, pure and applied, for the benefit of all humankind. In sustaining a peaceful world, prominent scientists from the last century, such as Lawrence Bragg, Kathleen Lonsdale, Linus Pauling, Dorothy Hodgkin and many others, made enormous contributions to world peace, both within unavoidable war and against avoidable wars. The United Nations and the United Nations Educational, Scientific and Cultural Organization International Science Years, most recently, the International Year of Crystallography and the International Year of Light, with their numerous activities, successfully brought scientific results to society at large and other research communities. The peaceful movement of scientists to labs for training and research collaboration should be encouraged. ICSU's free circulation of scientists statute must also be adhered to. These are examples of sustaining our contributions to research and societal challenges. I offer my own review and thoughts for my own research field of crystallography and sustainability in Reference 1. There are therefore many activities that you can take part in as a scientist to help with global sustainability and world peace.

So what would you wish to change in the organisation of global science?

REFERENCE

1. J. R. Helliwell (2015) Crystallography and sustainability. *American Crystallographic Association (ACA) Transactions Symposium 2015: Crystallography for Sustainability*, Volume 45. C. Lind-Kovacs and R. Rogers (eds.). pp. 8–19. Retrieved from http://www.amercrystalassn.org/documents/2015%20Transactions/volume45.pdf.

Section VIII

Appendices

In the appendices that follow, I now describe more generic skills.

Appendix 1: How to Deploy Some Very Basic Statistics When Necessary in Your Wider Roles as a Scientist

You may well find yourself on a committee, for example, for university staff safety or assessment of possible discrimination in university degree classifications for Black and other Minority Ethnic (BME) or gender groups, as I was, and where it will be assumed that as a scientist, you are literate in statistics. Hence, other committee members may well look to you for guidance on statistical matters. If they do not, you may well in any case need to butt into the discussion if it goes in the direction of insufficient data to prove their point.

I offer some thoughts in an appendix since this is a spinoff of your skills as a scientist. You will take on such roles to be a good general citizen in your department or university. Indeed, when called upon, you have to do your bit right!

In your research, you will be well trained in devising experiments to yield the data you need to prove or disprove your current scientific hypothesis. You will also wait until you have enough data. In a committee of the type I mentioned above, the data will not necessarily be presented in a useful form, nor enough of it. Let me provide an example for a safety committee monitoring the changes in the number of accidents in different categories; let us say worker accidents through lifting heavy boxes. The data presented are the last two years, namely 5 became 10 accident report forms. Aha! The cry goes up; the number of accidents has doubled. What should be done? At this point, you raise your hand and say that the numbers are too low to draw a conclusion yet. Monitoring for more years, by trawling back further in the record may reveal a trend. But, they say, the number has doubled! You then resort to the blackboard. Let us instead consider a case of 49 accidents in one year becoming 100 in the next. This again is doubling. But, and at risk of losing your audience, you explain the statistics:

Approximately, you explain, the fluctuation on a number is the square root of that number, that is, its standard deviation. The standard deviation on the difference between the two years is then

$$\left(\sigma_1^2 + \sigma_2^2\right)^{1/2} \tag{A1.1}$$

Hence, for the first case, the increase in the number of accidents is $(10 - 5) = 5$, with a standard deviation of 3.9. The difference 5 is therefore hardly more than one sigma!

In the second case, the increase in the number of accidents is $(100 - 49) = 51$ with a standard deviation of 12.2. The difference 51 is very significant at more than four sigma!

The second example that I will give you involves an average. These days, performance-ranking tables are everywhere. One such in the United Kingdom involved the performance of hospitals. I was sitting in my local café delicatessen one morning reading their preferred daily national newspaper. There was the head-line 'Half of UK's Hospitals Are below Average'. The paragraph of detail below the headline explained how terrible this state of affairs was. Now, of course, you have already realised that the moral of the story is that in such a list, there will indeed be half of the hospitals above average and a half below average. The state of affairs is only terrible if in examining the distribution of performance, the lower portion of the list has examples of hospitals falling below performance-required standards. This would never happen in science, you would say. Well, no. In my own field, the quality of protein crystal structures is presented with a slider diagram of different param-eters comparing a specific entry in the database with all other comparable crystal structures at that diffraction resolution. Indeed, whatever the actual numbers, which to be fair are quoted, the below average position nevertheless leads to that portion of the slider diagram entering red, that is, red = bad. If all the population of entries are in fact just fine, then the use of red in this way represents a needless anxiety for the depositors' files below average, and for the user. Obviously, if the population really does have poor quality entries, then these should be revealed or, better, not even deposited in the first place.

Appendix 2: How to Write Clearly and as Concisely as Possible

There is a long history of, and enthusiasm for, writing simply (and, likewise, in speaking or philosophising):

> If I'd had more time I would have written less.
>
> **Oscar Wilde**

> Je n'ai fait celle-ci plus longue que parce que je n'ai pas eu le loisir de la faire plus courte. [I have made this longer than usual because I have not had time to make it shorter.]
>
> **Blaise Pascal**

> If you want me to give you a two-hour presentation, I am ready today. If you want only a five-minute speech, it will take me two weeks to prepare.
>
> **Mark Twain**

> A scientific theory should be as simple as possible, but no simpler.
>
> **Albert Einstein**

> Pluralitas non est ponenda sine necessitate. [Plurality should not be posited without necessity.]
>
> **William of Ockham**
> *This principle being known as* Occam's razor

> The simplest explanation is usually the correct one.
>
> **Anonymous**

Writing clearly is a general skill. It neatly applies to how we write up our scientific results, as well as how we write anything, our e-mails, our letters etc. I decided to make this an appendix rather than a main chapter, but the importance of this topic and skill for a scientist remains.

Writing carefully requires time, and that of course is the essence of the quotations. With that said, being too short is in danger of ending up being cryptic, which detracts from getting your message across to a reader. That is the essence of Albert Einstein's remark. In today's computer word processing world, much help is available. Another aspect of clear writing is good syntax. Microsoft Word underlines a problem sentence with a wavy line. If you need a reference work on language syntax and grammar, there are various to choose from, for example, there is the older text *Manual of Composition and Rhetoric* [1]. Some publishers of scientific articles strive to help their authors with writing tools. One such is Publcif and PublBio of the IUCr Journals [2]. A book devoted to clear and effective writing for scientists is by Joshua Schimel [3].

Software tools in the end cannot catch the error that only a careful read can do. Let us take a practical example, your curriculum vitae (CV). You apply for a job; your future hopes rest upon your CV making a good impression. In it, you mention your hobbies as

Cooking my pets and travel.

It is quite different from

Cooking, my pets and travel.

You should also consult *Eats, Shoots and Leaves: The Zero Tolerance Approach to Punctuation* [4].

REFERENCES

1. J. H. Gardner, S. L. Arnold and G. L. Kittredge (1907) *Manual of Composition and Rhetoric*. Ginn and Company, Boston.
2. IUCr Journals (n.d.) Word templates and tools for Windows. Accessed 7 August 2016. Retrieved from http://journals.iucr.org/services/wordstyle.html.
3. J. Schimel (2012) *Writing Science: How to Write Papers That Get Cited and Proposals That Get Funded*. Oxford University Press, Oxford.
4. L. Truss (2011) *Eats, Shoots and Leaves: The Zero Tolerance Approach to Punctuation*. Fourth Estate, London.

Appendix 3: How to Keep to Budget

Keeping to the budget, like writing clearly, is a general skill. It neatly applies to how we live our lives as much as how we manage our scientific projects. I therefore also make this an appendix.

To keep to a project budget you need several things:

1. A good costing for your project in the first place, and even a simple check-list from your department finance team of what categories to cost, can be enormously helpful.
2. Regular budget statements from your department finance team or simple access tools to obtain the information you need as your project develops is also helpful.
3. In the scientific civil service, a contingency was an allowed line in one's instrument build or operational budget, which is simply good sense I think.

In the 1980s, I was presented with a printed computer output every month for each of my research project codes. I found this effective. In the 1990s, I could request budget progress reports. Since I found myself travelling a lot, and with the onus being on me to obtain the spend reports, it led me to not having such regular access to the project spend data I needed. So to avoid embarrassing overspend in specific categories, I also kept a notebook of what I had spent. In the 2000s, a web access tool to obtain one's project budget statements came about; this required formal training. I still found the procedure complex even after my training, which was also badly disrupted by unruly interruptions from a couple of angry participants, angry at the complexity of the new system. While there was something to be angry about the disruption was unhelpful to one's training.

The moral of the story is that unlike one's life at home where one might choose to write a letter to enquire about an account one has, rather than wait in an endless telephone enquiry queue, at work you are at the mercy of the system your institution weds itself to. Your only motto then I think is to grin and bear it, and start your own spend notebook of course.

Appendix 4: How to Observe

You see, but you do not observe. The distinction is clear.

Arthur Conan Doyle

In A Scandal in Bohemia in The Adventures of Sherlock Holmes (1892)

The art of seeing and listening, and the use of smell, can be combined to optimise your observations. Sherlock Holmes, in trying to understand a crime and who did it, carried his observations forward with his powers of deduction. There are several skills at work here to make a very powerful combination not only for the detective but also for the scientist.

In the laboratory context, I would quite often find myself asking how big was the crystal, what colour was the crystal or what temperature was it in the crystallisation room etc. These questions could be to a person or when I was reading a publication. The latter situation is quite startling but does happen. You read a whole publication, as have the authors, the referees and the editor who accepted it, and yet a core detail can be missing. Did all those people deliberately try and mislead the reader? Well, no. I very much doubt it! They, all of them, saw the words, but they simply just did not observe or deduce. Is this so different from Sherlock Holmes noticing that someone has a worn inside right trouser leg, which means that he or she is a cyclist? No. So my point is that your powers of observation are an important part of your life skills and your skills as a scientist.

How do you improve your powers of observation? One way is to imagine that you have been asked to describe a thief in the street or in a shop. Could you remember the details? What are the core details? Just try observing someone, and then close your eyes and try and recall them to describe them accurately.

How do you listen better? Try to recall a conversation. Do you only remember what you said? The art of listening involves, for example, not interrupting the other person. If they are exceptionally talkative, you may need to be assertive and say, 'What do you mean exactly?' You should also listen carefully for what is not said.

Appendix 5: How to Deal with Bullying

Bullying is a generic problem. You might encounter the bully at your local sports club or at your workplace. Unlike your sports club, for which you can decide to join somewhere else, the workplace bullying situation is not so simple to solve. The Times Higher Educational Supplement and Wikipedia [1] reports on surveys of the academic workplace and confirms that the problem exists in large numbers. However, these do not specify the science workplace in particular.

The general aspects of workplace bullying ranges into the following categories [2]:

- Threat to professional status, such as public professional humiliation, accusation of lack of effort and belittling of the victim
- Threat to social status, such as teasing and name-calling of the victim
- Isolation of the victim, such as withholding information and preventing access to opportunities, such as training workshops, attendance and deadlines
- Overwork of the victim, such as setting impossible deadlines and making unnecessary disruptions
- Destabilization of the victim, for example, setting meaningless tasks, not giving credit where credit is due and removal of the victim from positions of authority

The latter bullet point 'Removal of the victim from positions of authority' can occur if the bully has power, such as a head of a department, and can be construed as constructive dismissal.

Several aspects of academia lend themselves to the practice of bullying occurring and also discourage its reporting and mitigation. This is because the academic leadership is usually drawn from the ranks of faculty, most of whom have not received the management training that could enable an effective response to such situations. Thus, where freedom is possible, liberties by the bully will be taken. Academic victims of bullying may also be particularly conflict averse. The generally decentralised nature of academic institutions can make it difficult for victims to seek recourse, and an appeal to an outside authority is a dive into the unknown, at best. Therefore, caution and good independent advice are needed before reporting any problems or lodging a formal complaint. The academic workplace does have a union for its academic employees, or should have; in the United Kingdom it is the University Colleges Union and which provides helpful counsel on such problems as a benefit of membership.

The best starting point though to deal with a bully in my view is to say directly to them, 'Why do I feel bullied by you?' or 'Please stop doing this'. There are also courses on assertiveness training in group discussions to help you if you are shy.

While not directly aiming at the bully problem, clear and confident speaking, especially if you are a person who is conflict averse, can be of help to you before launching into an official complaints procedure.

REFERENCES

1. Wikipedia (n.d.) Workplace bullying in academia. Accessed 7 August 2016. Retrieved from https://en.wikipedia.org/wiki/Workplace_bullying_in_academia.
2. K. Rigby (2002) *New Perspectives on Bullying*. Jessica Kingsley Publishers, London.

Appendix 6: How to Take Decisions

On a wrong decision:

> Anyone who has never made a mistake has never tried anything new.
>
> **Albert Einstein**

On taking many wrong decisions:

> I have not failed. I have just found 10,000 ways that won't work.
>
> **Thomas Edison**

On indecision:

> When you come to a fork in the road, take it.
>
> **Yogi Berra**
> *USA professional baseball catcher, manager and coach*

There are decisions that we face in our home and in our scientific life. Both aspects of our life can include operational, day-to-day decisions or tactical medium term decisions relating to an overall long-term strategy and the long-term strategic decisions themselves that guide our life or our laboratory.

At home, we might well say to our partner that we will have to make a decision about a new car or a new laptop. We can go and buy a magazine: Which car? Or Which laptop? And these can be very helpful in matching what is available to what one's needs are.

At the lab, we may say that we will have to decide between this supplier or that supplier of a microscope, a diffractometer, an electron microscope or a nuclear magnetic resonance spectrometer. The companies will provide glossy leaflets along with a sales representative visit by arrangement and even a visit somewhere else for you to make your own tests of their apparatus. As well, we can maybe e-mail a bulletin board for information, to be e-mailed back in reply, on list or off list, by helpful colleagues. Or attendance to a conference commercial exhibition can be highly informative of course.

In government, a parliament of elected representatives may address the decision for where shall we put our new synchrotron or neutron source.

Some decisions may not be too crucial, such as whether we turn left or right; we can turn around later if needs be, but we cannot just sit in the car at the road junction in a state of indecision; that is the Yogi Berra wisdom!

There are many good how-to guides as to how to take a decision. These go along the lines of: identifying a need, gathering information and data, evaluating the pros and the cons or the cost and the benefits of each option, and taking a decision. The final decision can be by a person (most likely the person who signs the cheque or the risk holder) or a group. The group may have to operate within a constitution of needing a unanimous vote or, easier, a majority vote for less important or less costly decisions than those requiring a unanimous vote. Also, sometimes, there is no good option, only the least worst option, and you decide to take that option simply because anything else would be far worse. These guides also encourage the writing down of the pros and the cons of not taking a decision, which is also good advice originally written about by Benjamin Franklin in 1772 (cited in Reference 1).

In the end, for a decision taken but regretted, a postmortem analysis may become necessary for future reference as a learning experience, which is also good practice. As Einstein's and Edison's quotes illustrate, decisions involve risk, and therefore, mistakes can and will occur. But as scientists, we accept probability and how to quantify risks. And sometimes, we like to innovate and be adventurous, with risk, and at other times, we proceed not by leaps but incrementally with low or zero risk. What we do value is proceeding when we have all the relevant information that can be had at the time.

Although, overall, we as scientists can see or listen to bad decisions that our colleagues have taken! 'How so?', you ask.

Have you not read a publication where there are no conclusions? That paper was the result of a bad decision by the authors of that paper to submit it and then also by the referees and editor who passed it through. If your research has not yet reached any conclusions, why would you publish?

Have you listened on the radio or read in a newspaper about a press release of some scientific results but where the implications are so hedged in with 'We cannot be certain that this explains … or that this work may lead to a new cure but maybe in five years there is hope that it may'. If the person speaking about the press release has to say that, why have the coauthors in the research allowed a press release at all? Is such a premature press release causing damage in fact to science? Or is it the need, these days by way of justification, to warrant such a press release? I am not at all sure that such premature trumpeting is a good idea at all for us as a community. My friends at my tennis club are not impressed, nor my relatives, when I witness their reactions to such press release stories.

One of my most unexpectedly successful press releases concerned the lobster coloration results (see Chapter 25). The public and the schoolchildren were simply intrigued by the colour change effect! We had a portion of the press release text that mentioned the health benefits of the carotenoid at the centre of the colour change effect, but I am not so sure that it was a good decision to include that. Actually, it was harmless enough deciding to include that applied aspect as people mainly focused on the colour change mystery bit of the story! An interesting summary of issues, pros and cons, also including social science research, in presenting research results to the public and preparing for possible antagonists to your message is to be found in Chapter 8 of Badgett's book [2].

Overall, this is a big topic, but I hope these observations help you when you face a decision. Oh yes, and do not forget, delay can occasionally be the best option as I described in Chapter 21, even if you might be labelled a *procrastinator*.

REFERENCES

1. D. G. Ullman (2006) *Making Robust Decisions: Decision Management for Technical, Business & Service Teams.* Trafford Publishing, Bloomington, North America.
2. M. V. Lee Badgett (2016) *The Public Professor: How to Use Your Research to Change the World.* New York University, New York.

Appendix 7: How to Be a Project Sponsor

So by the time I had reached my early fifties, I had extensive experience of running scientific research projects including very large ones with multimillion-pound budgets, knowing about risk management, understanding how to benchmark the instrumentation I had either developed myself or relied on so I knew it was working smoothly, knowing a lot about computation and data analysis as well as the necessary hardware, keeping good records, hiring staff and so on. So one day, the university registrar rang me up and said, 'John, we need someone to be Project Sponsor of the new Student System Project because the person who originally took it on we have had to ask to take on something else. We know you have had a lot of experience running big research projects as well as computing. Would you take it on?' After the usual discussions (what does it involve, how much time do you think it would take etc), and learning that the core point with the new system was for the student records to become web enabled for data entry and access, I said yes.

A project sponsor is a pivotal role in a project where one represents the business (the university) to the project team (information technology [IT] specialists in this case), and one represents the project to the university, the senior executive and ultimately, on a weekly even daily basis, to one's colleagues. One regularly interacts with the overall project manager.

Only later did I reflect on the fact that I should have asked for a pay raise.

More than 10 years later, the implemented system is still in place. So in spite of the fact that there did seem to be quite a few public high-profile IT system fails or 'in deep trouble' (examples in the United Kingdom: car tax office records, passport office records, income tax office records, and biggest of them all, patient healthcare records) that made me feel uncomfortable while listening to the BBC Radio 4 news, ours is a fully fledged, still working, web-enabled system for secure and accurate student data entry and access.

Appendix 8: How to Write a Reference

You may often need to write a reference, mainly in your scientific work but also in other circumstances. The nature of writing these is evolving as the rights of the individual being written about to see what is being written are becoming more common. The institution may well be from a different country than your own and where the rules are different from your country's rules. So in order to be fair to both the institution and the applicant, and be objective, I work to the following principles:

1. Be evidenced based in what you write.
2. Do include openly available publication metrics or, in the case of students, quote exam marks (with the student's permission), as a part of your reference and to assist you in the writing of your conclusions on the overall suitability of the candidate.
3. Do use the tabulation of specific qualifications and skills required for the post to offer your evidenced-based observations in the way that is structured to the preferences of the organisation requesting your reference.
4. Avoid giving an opinion and instead concentrate on giving objective guidance on whether the experience of the person matches the job specification.

Unless I am specifically instructed not to do it, I request the person I am writing about to check that the evidence I am citing is accurate and reasonably complete.

Appendix 9: How to Delegate

If you want something done right, do it yourself.

A well-known saying

The quote advises against delegation. But at home, we regularly delegate tasks. As an example, we might like to decorate the house ourselves to ensure that in every detail it will be as we would wish to have it. But if we are busy at work, or incapacitated in some way, we need to get a decorator to do it. This contract is a moment of delegation. Having agreed to the work and the price and when it will be started and finished, you give the agreement to the decorator to start the work. It is unlikely that there will need to be any deviation from the agreed work, but a starting to rot window frame will bring about a discussion about a new one, what type, what price and who would fit it.

In Chapter 6, I referred to the art of delegation in the following way: Do not micromanage your team; you have appointed them because they are good at what they do. Basically, you set an overall goal or strategy for your staff, but give them the freedom to reach the objectives in their own way.

In Chapter 2, I explained the useful definitions of *smart* objectives, namely

- Specific – Target a specific area for improvement.
- Measurable – Quantify or at least suggest an indicator of progress.
- Assignable – Specify who will do it.
- Realistic – State what results can realistically be achieved, given the available resources.
- Time-related – Specify when the result(s) can be achieved.

An interesting, modern and developing area of delegation is in the harnessing of robots. In my own research field, this increasingly involves the use of robotic devices for the systematic and precise exploration of the chemical parameters, of which there are many, for crystallising a protein. Before automation, this was undertaken solely by people and inevitably restricted the number of crystallisation experiments and also led to occasional errors of volume quantities and weights of the chemicals used. At worst, researchers reading a publication might find that they could not repeat the crystallisation. Fortunately, this to my experience would be in rare cases, although statistics on this irreproducibility are naturally not available. This area then, the reproducibility of experiments, I think we can assume, has improved in terms of the certainty of the chemical conditions reported by the use of robotics. Robotics has also made a major impact, again an example from my field of research, in the placement of a protein crystal in a synchrotron X-ray beam. This has greatly improved the throughput of firstly testing the crystals for which one diffracts the best. Then, as the X-ray beam intensities have gotten stronger and stronger, the time spent collecting each dataset is now very short (fractions of a second versus hours when I started out at my national synchrotron, the SRS), so that a robot changing a sample is much faster than when I was doing the step manually. Initially the robots were blindly doing this, whereas I could see that a crystal might have interesting flaws like being a twin. So for the robot, we have now added a high-resolution camera, and pictures of each crystal are also recorded! Also, overall, robots do not get tired and can work round the clock! A very impressive example is described in Ref. [1].

So whether with a robot or a team of people there are various circumstances where one needs to delegate objectives, and it is much better than always retaining the tasks to oneself. I have provided you with some illustrations and the core principles regarding the delegation of smart objectives.

REFERENCE

1. M. W. Bowler, O. Svensson and D. Nurizzo (2016) Fully automatic macromolecular crystallography: The impact of MASSIF-1 on the optimum acquisition and quality of data. *Crystallography Reviews*. Open access review article. doi: 10.1080/0889311X.2016 .1155050.

Appendix 10: How to Prepare to Be a Science Expert Witness

The invitation to me to be a science expert witness in a case about a pharmaceutical, involving the uniqueness of powder X-ray diffraction evidence of a new drug polymorph, came to me by e-mail. Due to a diary clash, I could not take up the invitation, but my curiosity about presenting science in the courtroom was aroused. I consulted a colleague in my research field, Professor Mike Glazer, who I knew had served in this role, and I later invited him to review a very relevant book [1], whose book review was duly published [2]. I am also a keen watcher of courtroom dramas such as the *Law and Order* TV episodes, especially the series set in New York, which neatly captured various important social as well as occasionally scientific evidence-driven cases. In the United Kingdom, we also have our own TV courtroom dramas with favourites such as Rumpole of the Old Bailey, whose most famous case was the Penge Bungalow Murders and where the evidence of the blood proved critical. More generally though, how does science bear up under a legal scrutiny?

I continued to learn more about the scientist as expert witness in the court room. There is the US National Academies Press (NAP) handbook [3], whose contents are described at the NAP's website as follows:

> The *Reference Manual on Scientific Evidence*, Third Edition, assists judges in managing cases involving complex scientific and technical evidence by describing the basic tenets of key scientific fields from which legal evidence is typically derived and by providing examples of cases in which that evidence has been used.
> First published in 1994 by the Federal Judicial Center, the Reference Manual on Scientific Evidence has been relied upon in the legal and academic communities and is often cited by various courts and others. Judges faced with disputes over the admissibility of scientific and technical evidence refer to the manual to help them better understand and evaluate the relevance, reliability and usefulness of the evidence being proffered. The manual is not intended to tell judges what is good science and what is not. Instead, it serves to help judges identify issues on which experts are likely to differ and to guide the inquiry of the court in seeking an informed resolution of the conflict.
> The core of the manual consists of a series of chapters (reference guides) on various scientific topics, each authored by an expert in that field. The topics have been chosen by an oversight committee because of their complexity and frequency in litigation. Each chapter is intended to provide a general overview of the topic in lay terms, identifying issues that will be useful to judges and others in the legal profession. They are written for a non-technical audience and are not intended as exhaustive presentations of the

topic. Rather, the chapters seek to provide judges with the basic information in an area of science, to allow them to have an informed conversation with the experts and attorneys.

Chapter 13 entitled 'On the Witness Stand' in Cornelia Dean's book [4], includes a description of the tensions involved. These include contrasting matters such as 'Scientists and engineers start with questions and look for answers'. '... By contrast, the law works on the premise that the best way to learn in a dispute is to have each side advance the strongest arguments it can to make its case. Lawyers on each side know their desired outcome ... and are not obliged to produce evidence that hurts their case'. This leads on to the problem of the scientist as the hired gun for one side versus the other side's own scientist hired gun! Of course, the role of say a pathologist at a coroner's inquest into a death can be a matter-of-fact presentation of the scientific evidence that is necessarily, and legally bound to be, complete.

Overall, in many US states, for example, Rule 702 governs testimony by expert witnesses namely

A witness who is qualified as an expert by knowledge, skill, experience, training, or education may testify in the form of an opinion or otherwise if:

(a) The expert's scientific, technical, or other specialized knowledge will help the trier of fact to understand the evidence or to determine a fact in issue;
(b) The testimony is based on sufficient facts or data;
(c) The testimony is the product of reliable principles and methods; and
(d) The expert has reliably applied the principles and methods to the facts of the case.

So where do I think we stand in looking at an invitation to serve as a scientific expert witness? You obviously, first, have to believe in the evidence you are invited to speak to. Secondly, you must be qualified, as defined in Rule 702. Thirdly, there are different types of expert witness, namely whether you would be adviser to one or the other side of the case protected therefore by attorney privilege or whether you would be a testifying expert witness or not and finally appointed by the court; these roles and the legal procedures under which you would be bound are carefully explained by James Speight [1].

Clearly, this is not a commission to be entered into lightly!

REFERENCES

1. J. G. Speight (2008) *The Scientist or Engineer as an Expert Witness*. CRC Press, Taylor & Francis, Abingdon.
2. A. M. Glazer (2011) Book review of *The Scientist or Engineer as an Expert Witness Crystallography Reviews* 17: 145–146.
3. Committee on the Development of the Third Edition of the Reference Manual on Scientific Evidence; Committee on Science, Technology, and Law; Policy and Global Affairs; Federal Judicial Center; National Research Council (2011) *Reference Manual on Scientific Evidence*, third edition. National Academies of Science Press, Washington, DC.
4. C. Dean (2009) *Am I Making Myself Clear? A Scientist's Guide to Talking to the Public*. Harvard University Press, Harvard, MA.

Appendix 11: How to Explain the Scientific Method to the Public and Schoolchildren

It is likely that you have asked yourself at some stage what the scientific method is exactly. Maybe, this was when you were preparing your first research grant proposal. These can be of different types. They therefore illustrate the different types of science or science investigation, which are your own lab's various examples of the scientific method. You may have a starting hypothesis or theory and you wish to prove it. Or you know of an important topic, and there is a collection of facts about it, data, out there, but whose data set is incomplete, and you wish to extend it to something more complete and useable. Or finally, you have an idea for a new method that you think could be important to collect new scientific data quicker or better or both, and you wish to validate your proposed new method. All these examples can be featured in a science research grant proposal.

The time will also surely come when someone will look to you as a scientist to explain as simply as possible the scientific method. This request may well arise quite probably from a member of the public or a school pupil, and so your explanation will need to be especially clear. For this, you will need a few everyday examples. As the opening paragraph states, you have a suite of your own examples but probably imbued with so much technical jargon to render them unusable as examples. So you have to become the general philosopher of science and harness some well-known examples. A complete introduction to the topic by a professional philosopher of science is by A. F. Chalmers [1]. I have already touched upon the philosophy of science when I explained the nature of scientific revolutions and the paradigm shifts explained by Thomas Kuhn [2] in Chapter 3.

At the outset, you could explain to the person who asked you that experimental science is based on facts. We see what the facts are, and we can gather these into a theory or a law and predict other facts that we might set about looking for from the first set of facts. Alternatively, you could explain that sometimes, driven by a scientific theory, the scientist goes on a quest for specific facts to verify the theory.

So which examples will you pick to illustrate the strengths of experimental science and to illustrate the strengths of theoretical science?

In the case of theories, there are basically two approaches; one is to prove a theory true, and the other is to prove a theory false. The best theory as viewed by Sir Karl Popper (1902–1994) is the one that has been tested the most often and has not been refuted! Popper used the black swan analogy. No matter how many white swans you encounter, you never know for sure that all swans are white, and only when you encounter a black swan can you successfully refute the theory that all swans are white. Also, Popper stressed that for a theory to be falsifiable, it has to be as specific as possible in its assertions.

So let us move to practical, well-known examples.

Firstly, consider Charles Darwin on the voyage of the Beagle in the 1830s. He looked at the wildlife of the geographically isolated Galapagos Islands and the types of bird known as finches in particular. The collections he made led to his theory of evolution based on natural selection, the survival of the fittest. This theory has not been refuted.

Secondly, let us consider Albert Einstein's theory that matter and energy were interchangeable. He derived his famous equation relating the energy, E, the mass, m, with the speed of light, c, which is $E = mc^2$. This was experimentally verified in the process known as *nuclear fission*, the breakup of an atomic nucleus upon absorption of a neutron for example involving the uranium isotope 235. This led onto predicting that a fission nuclear bomb could be made if a critical mass of the correct weight of two portions of that specific uranium isotope, one that was close to instability, could be suddenly brought together. This was of course experimentally proven subsequently.

A third example is that theories exist and seem to apply even where contradictory. The nature of light is a famous example here. In some experimental circumstances, such as diffraction and interference, light is best described by a wave theory, and in others, such as the photoelectric effect, light is best described as a photon.

There is a fourth situation I think, namely where intuition comes into play. Here one has sufficient experimental and theoretical experiences as to sense that one might discover something new if you investigate a new phenomenon. How to define intuition? It is more than feeling but not so clear as logical thinking. We can also call it a *hunch*. I shared some of my own hunches early on in this book in Chapter 2 'How to Recognise a Good Idea'.

While those mentioned earlier are good examples of the scientific method in practice to refer to because the public are likely to have heard of at least one of them, one could add that an often-used approach as a scientific method, step by step, is as follows:

1. Decide which observations to make.
2. Make those observations.
3. Create a hypothesis.
4. Predict outcomes.
5. Test the predictions (and hence the hypothesis) with new observations.

The National Physical Laboratory based at Teddington, UK, has an extensive, helpful website on a wide range of scientific topics; I showed as an example their outreach card explaining the scientific method in my introduction.

REFERENCES

1. A. F. Chalmers (1999) *What Is This Thing Called Science?*, third edition. Open University Press, Buckingham.
2. T. S. Kuhn (1996) *The Structure of Scientific Revolutions*, third edition. The University of Chicago Press, Chicago.

Bibliography

Science:
Peter B. Medawar (1979), *Advice to a Young Scientist*, Penguin Books, London.
Edward O. Wilson (2013), *Letters to a Young Scientist*, Live Right Publishing Corporation, W W Norton and Co., New York.

Science career guide:
Finlay MacRitchie (2011), *Scientific Research as a Career*, CRC Press, Boca Raton, FL, USA.

Concerning the academic life:
Joëlle Fanghanel (2011), *Being an Academic*, Routledge, Abingdon, UK.

Concerning journals:
Irene Hames (2007), *Peer Review and Manuscript Management in Scientific Journals: Guidelines for Good Practice*, Wiley-Blackwell, Malden, USA.

Concerning philosophy and ethics:
Erich Fromm (1947), *Man for Himself: An Enquiry into the Psychology of ethics*, Owl Books, New York.

Classic 'How to' books (politics):
Gerald Kaufmann (1997), *How to be a Minister*, Faber & Faber, London.
Paul Flynn (2012), *How to be an MP*, Biteback, London.

Name Index

Subject Index

Strengths Weaknesses Opportunities Threats
(SWOT) analyses, 47–50, 106,
134–135
Student Disciplinary Committees, 28
Student transfer report, 28
Supervision of postgraduates, 28, 30, 99–103
Sustainability, 113, 156, 162
Swinburne University of Technology 66
Swiss-Prot, 20
SWOT analyses, *see* Strengths Weaknesses
Opportunities Threats (SWOT) analyses
Synchrotron-Light for Experimental Science
and Applications in the Middle East
(SESAME), 162
Synchrotron Radiation Source, xiii, 13, 23,
47–48, 119

TARDIS, Australian synchrotron data store, 67
Taxpayers, 70, 76
Taylor and Francis Publisher, xiv, 88, 149–150
Teaching undergraduates, 29–30, 95–98, 130
Technology, 119, 124, 159
Television (TV), 112, 185
Temperament, 90, 147
Tenure, obtaining, 5, 105–108
Time management, 19, 43–45, 106
Transcript of marks achieved by student, 95
Tutorials, 30, 96, 98–99
Twitter, 37, 49, 51–53

Unconscious gender bias, 121
Undergraduates, 29–30, 95–98, 130
Unfunded research, 38, 49, 70, 86, 160, 161
United Kingdom Parliamentary Select
Committee on Science and
Technology, 159–160

United States Atomic Energy Commission,
146
University of Manchester, xi, xii, 6–7, 19, 23,
27–28, 30–31, 41, 52–53, 56, 86,
95, 105, 113–119, 121–124, 137,
148, 153
UNESCO, United Nations Educational, Scientific
and Cultural Organization, xiv, 85,
122–126, 162
UN, United Nations, 85, 162

Vision, 6, 60, 66, 69–72, 85, 87, 129, 152

Wellcome Trust, The, xiii, 20, 71, 75, 100
Wider roles as a scientist, 165
WiSET, Women in Science, Engineering and
Technology, 124
Women, boardroom, 123
Work-life balance, 45
Workshops
capacity building, 155
early career scientist, 105
research, 67
skills for continued professional development,
100, 134
teaching, 96, 98
World peace, how scientists can contribute to,
162
Writing
clearly and concisely, 149, 167–168
a reference, 181–182

X-ray crystallography, 11, 13, 16, 31, 36, 40,
56–57, 82, 98, 102, 113, 119

York University, 1–2, 6, 47–48, 56